Anas El Azizi El Alaoui
Abdelali Fateh

# Urban resilience in Morocco in the face of of globalization

Anas El Azizi El Alaoui
Abdelali Fateh

# Urban resilience in Morocco in the face of of globalization

## Adaptation and sustainable development strategies for Moroccan cities

ScienciaScripts

**Imprint**

Any brand names and product names mentioned in this book are subject to trademark, brand or patent protection and are trademarks or registered trademarks of their respective holders. The use of brand names, product names, common names, trade names, product descriptions etc. even without a particular marking in this work is in no way to be construed to mean that such names may be regarded as unrestricted in respect of trademark and brand protection legislation and could thus be used by anyone.

Cover image: www.ingimage.com

This book is a translation from the original published under ISBN 978-620-6-71657-0.

Publisher:
Sciencia Scripts
is a trademark of
Dodo Books Indian Ocean Ltd. and OmniScriptum S.R.L publishing group

120 High Road, East Finchley, London, N2 9ED, United Kingdom
Str. Armeneasca 28/1, office 1, Chisinau MD-2012, Republic of Moldova, Europe
Printed at: see last page
**ISBN: 978-620-8-02653-0**

# *Urban resilience in Morocco to the challenges of globalization*

*Adaptation and sustainable development strategies for Moroccan cities*


Anas EL AZIZI EL ALAOUI
Abdelali FATEH

# *Foreword*

Urban resilience has become a central concept in the study and management of contemporary cities, particularly in the context of the accelerated globalization that characterizes our times. This multi-dimensional paradigm provides a relevant framework for analysis and action to address the complex challenges facing urban agglomerations in the 21st century. Morocco, a country in the throes of socio-economic and territorial change, is an ideal case study for exploring the dynamics of urban resilience in the face of the multifaceted pressures of globalization. This book takes an in-depth look at the issues, constraints and opportunities involved in strengthening the resilience of Moroccan cities in a constantly evolving, interconnected world.

Economic globalization, with its flows of capital, goods and information, has profoundly reconfigured urban spaces on a planetary scale. Moroccan cities are no exception to these transformative dynamics which, while offering new prospects for development, also generate heightened tensions and vulnerabilities. Economic restructuring, demographic pressures, socio-spatial inequalities and environmental challenges are all factors that test the capacity of urban systems to adapt and reinvent themselves. In this context, urban resilience emerges as an operative concept for articulating the economic, social, environmental and institutional dimensions of sustainable urban development.

This book offers a multiscalar, transdisciplinary analysis of urban resilience in Morocco, drawing on the most recent theoretical and methodological contributions to this rapidly expanding field of research. It aims to decipher the complex mechanisms underlying the capacity of Moroccan cities to absorb shocks, adapt to change and transform in the face of the challenges of globalization. Using a systemic approach, it explores the interactions between the various components of urban resilience and the strategies implemented by local players to enhance the robustness and flexibility of urban systems.

The ambition of this work is twofold: on the one hand, to contribute to the advancement of scientific knowledge on urban resilience in the specific context of emerging countries and, on the other, to provide relevant insights to guide public action and decision-making on sustainable urban development in Morocco. Combining analytical rigor and operational scope, this book is aimed at researchers and students of urban studies, as well as practitioners and decision-makers involved in the governance and development of Moroccan cities.

*Anas EL AZIZI EL ALAOUI*
*Faculty of Arts and Humanities*
*Mohammed V University, Rabat - Morocco*

# Table of contents

## *General introduction: Contextualizing urban resilience in a world of accelerating change*

In today's fast-changing world, urban resilience is an essential paradigm for understanding the complex challenges facing contemporary cities. Globalization, characterized by the intensification of flows of capital, people, technology and information on a planetary scale, has profoundly reconfigured urban dynamics, generating both development opportunities and heightened vulnerabilities. This dialectic between global integration and local fragility places cities at the heart of the challenges of sustainability and adaptation to the multifaceted changes that characterize the Anthropocene. In this context, urban resilience is emerging as an operative concept for articulating the economic, social, environmental and institutional dimensions of urban development, while taking account of territorial specificities and the capacities for action of local players.

The acceleration of global changes, whether technological, economic, climatic or societal, is putting unprecedented pressure on urban systems, testing their capacity to absorb shocks, adapt to change and transform in the face of emerging challenges. Moroccan cities, like many agglomerations in emerging countries, face a constellation of interconnected challenges: economic restructuring and deindustrialization, demographic pressures and migration, growing socio-spatial inequalities, environmental degradation and natural hazards exacerbated by climate change. These multiscalar and multidimensional dynamics call for a holistic and systemic approach to urban resilience, capable of grasping the complexity of interactions between the various components of urban ecosystems and their overall environment.

The conceptualization of urban resilience in this context of accelerated globalization requires us to move beyond sectoral and static approaches, in favor of a dynamic and integrative vision of urban development. The aim is to understand how cities, as complex socio-ecological systems, can not only maintain their essential functions in the face of disruption, but also reinvent themselves and evolve positively in the face of the structural changes brought about by globalization. This perspective involves a detailed analysis of the adaptive capacities of cities, their multi-level governance mechanisms, and the social, technological and institutional innovations emerging in response to contemporary challenges.

In the specific case of Morocco, the study of urban resilience takes place in a context of multiple transitions: demographic transition, economic transition towards a service and knowledge economy, ecological transition in the face of sustainability imperatives, and political transition towards more participative and decentralized governance. These transformations, which are taking place on different scales and at different speeds, provide fertile ground for experimenting with new approaches to resilient urban development. An analysis of the strategies implemented by Moroccan cities to strengthen their resilience in the face of the challenges of globalization thus sheds valuable light on the dynamics of urban adaptation and innovation in an emerging country context.

By adopting a critical and analytical perspective, this study aims to deconstruct the dominant discourses on urban resilience and propose a contextualized and nuanced reading, attentive to the specificities of the Moroccan terrain. It explores the tensions and synergies between the different dimensions of urban resilience, as well as the issues of power and equity that underpin resilience-building processes. This approach sheds light on the methodological and conceptual challenges involved in operationalizing urban resilience in a context of globalization, while opening up avenues of reflection for more adaptive and inclusive urban governance in the face of the uncertainties of the contemporary world.

## *Chapter 1: Urban resilience -Theoretical conceptualization and operationalization of a multidimensional paradigm*

The concept of urban resilience has established itself as a central paradigm in contemporary urban studies, offering a theoretical and operational framework for understanding the complexity of urban systems in the face of the multifaceted challenges of the 21st century.

This emergence takes place against a backdrop of growing awareness of the vulnerability of cities to exogenous shocks and endogenous pressures, exacerbated by the dynamics of globalization and climate change. This chapter explores in depth the theoretical foundations and practical implications of urban resilience, highlighting its multidimensional nature and its relevance to the analysis of contemporary urban transformations.

The conceptualization of urban resilience has its roots in a variety of disciplinary fields, from ecology to psychology to complex systems science. This conceptual hybridization lends the notion of urban resilience a heuristic richness, but also raises epistemological challenges in terms of its definition and operationalization in the specific context of urban studies.

A critical analysis of the various interpretations

and applications of urban resilience reveals the need for an integrative approach, capable of capturing the dynamic interactions between the economic, social, environmental and institutional dimensions of urban systems.

This chapter will focus on deconstructing the components of urban resilience, examining how this concept relates to other key notions such as sustainability, adaptability and transformability of cities. It will also explore the methodological and practical implications of adopting a resilience perspective in urban planning and governance.

Adopting an analytical and critical approach, we will examine the potential and limits of this paradigm for rethinking urban development in a context of accelerating change and growing uncertainty.

The aim of this chapter is twofold: firstly, to provide a sound conceptual framework for the analysis of urban resilience in Morocco, and secondly, to contribute to the scientific debate on the operationalization of this concept in a variety of urban contexts.

By examining the different dimensions of urban resilience and their interactions, we will lay the theoretical foundations for an in-depth understanding of the issues and strategies for building resilience in Moroccan cities in the face of the challenges of globalization.

## 1.1 The emergence of resilience as a key concept in

**urban studies**

## 1.1.1 Interdisciplinary origins of the concept of resilience

The concept of resilience, in its trajectory towards application to urban systems, finds its roots in a rich interdisciplinary soil, testifying to the complexity and transversality of the issues it seeks to apprehend. This plural conceptual genealogy gives the notion of urban resilience remarkable analytical depth, while posing significant epistemological challenges to its operationalization in the specific field of urban studies. Exploring these interdisciplinary origins is therefore crucial to understanding the scope and limits of the resilience paradigm as applied to contemporary cities.

Contributions from ecology and the sciences of complex systems form the fundamental foundation on which the notion of urban resilience has been built. Crawford Stanley Holling's (1973, p. 14) pioneering work in ecology introduced a conception of resilience as "a measure of the persistence of systems and their capacity to absorb change and disturbance while maintaining the same relationships between populations or state variables". This approach, initially applied to ecosystems, paved the way for a dynamic, non-linear understanding of complex systems, challenging stable equilibrium paradigms in favor of a vision of multiple equilibria and adaptive cycles. The transposition of these concepts to urban systems, conceptualized as complex socio-ecosystems, has considerably enriched the analysis of urban dynamics in the face of disturbance and change (Cyril Folke, 2006, p. 258259).

At the same time, the social sciences, and psychology in particular, have added a further dimension to the conceptualization of resilience, emphasizing the adaptive and transformative capacities of individuals and communities in the face of adversity. In particular, Boris Cyrulnik's (2001, p. 45) work on psychological resilience has helped to broaden understanding of the concept beyond mere resistance, to include notions of rebound and positive development despite trauma. This psychosocial perspective has found particular resonance in the analysis of urban resilience, highlighting the importance of social, cultural and institutional factors in the ability of cities to cope with crises and reinvent themselves.

The integration of these different approaches into the field of urban studies has taken place gradually, catalyzed by the growing recognition of the vulnerability of cities to global challenges such as climate change, economic crises or pandemics. In particular, the work of Serge Lhomme, Damien Serre, Richard Laganier and Youssef Diab (2010, p. 172) has helped formalize a systemic approach to urban resilience, proposing an analytical framework that integrates the technical, organizational and social dimensions of urban systems. This interdisciplinary convergence has considerably enriched our understanding of resilience mechanisms at city scale, while raising new questions about the measurement and operationalization of this multidimensional concept.

Nevertheless, this conceptual hybridization is not without its epistemological and methodological challenges. The polysemy of the term "resilience" and the diversity of its interpretations across disciplines raise the question of the coherence and specificity of its application to urban systems. As Magali Reghezza-Zitt (2013, p. 35) points out, "resilience appears to be a nomadic concept, whose semantic plasticity allows for a variety of, sometimes contradictory, uses". While this conceptual flexibility offers undeniable heuristic richness, it also calls for greater vigilance in defining and operationalizing urban resilience, to avoid the pitfalls of over-extensive or imprecise use of the concept.

In conclusion, the interdisciplinary origins of the concept of resilience are both a strength and a challenge for its application in the field of urban studies. This multi-faceted genealogy offers a rich conceptual framework for understanding the complexity of urban systems in the face of the multifaceted disturbances of the contemporary world. However, it also imposes analytical rigor in the transposition and adaptation of these theoretical contributions to the specificities of urban contexts, opening up new perspectives for research and action to strengthen the resilience of cities in the face of the challenges of the 21st century.

## 1.1.2 Historical evolution of the concept's application to urban studies

The historical evolution of the application of the concept of resilience to urban studies bears witness to a complex process of adaptation and conceptual enrichment, reflecting the profound changes in urban development paradigms in the face of the emerging challenges of the contemporary world. This trajectory, far from being linear, has been characterized by a series of theoretical and practical transitions and reorientations, marking the shift from a vulnerability-centric approach to a more holistic and dynamic vision of urban resilience.

In the 1990s and early 2000s, urban studies were largely dominated by the vulnerability paradigm, focused on identifying and reducing the fragility of urban systems in the face of natural and man-made hazards. As Patrick Pigeon (2005, p. 27) points out, "vulnerability appeared to be the key concept for understanding and managing urban risks, emphasizing the structural and organizational weaknesses of cities". This approach, while relevant for identifying the weak points of urban systems, often proved insufficient for understanding the adaptive and transformative capacities of cities in the face of disruption.

The transition to the resilience paradigm in urban studies began gradually, catalyzed by a growing awareness of the complexity and interconnectedness of contemporary urban challenges. The work of Cyril

Folke, Steve Carpenter, Thomas Elmqvist, Lance Gunderson, Crawford Stanley Holling and Brian Walker (2002, p. 437) played a crucial role in this evolution, proposing a conceptualization of socio-ecological resilience that emphasized "the capacity of a system to absorb disturbance and reorganize itself while undergoing change, so as to retain essentially the same function, structure, identity and feedbacks". This approach opened the way to a more dynamic and multidimensional understanding of urban resilience.

The gradual integration of resilience into urban policies and planning accelerated in the second half of the 2000s, driven by major natural disasters and global economic crises that highlighted the vulnerability of urban systems to exogenous shocks. The "Making Cities Resilient" campaign launched by the United Nations in 2010 marked a significant turning point in this dynamic, placing resilience at the heart of international urban agendas. As Géraldine Djament-Tran, Antoine Le Blanc, Serge Lhomme, Samuel Rufat and Magali Reghezza-Zitt (2011, p. 11) analyze, "this institutionalization of urban resilience has helped broaden the concept's scope beyond risk management alone, to encompass broader issues of sustainable urban development".

This conceptual evolution has been accompanied by a significant methodological enrichment, with the development of tools and indicators aimed at operationalizing urban resilience. In particular, the work of Damien Serre, Bruno Barroca and Richard

Laganier (2013, p. 56) has helped formalize approaches to assessing urban resilience, proposing analytical frameworks that integrate the technical, organizational and social dimensions of urban systems. This methodological evolution has enabled us to move beyond purely descriptive approaches towards more prescriptive and operational models of urban resilience.

At the same time, the emergence of "smart cities" and the acceleration of the digital revolution have opened up new perspectives for applying the concept of resilience to urban studies. As Antoine Picon (2018, p. 83) points out, "the integration of digital technologies into urban management has reconfigured the ways in which we understand and act on city resilience, offering new levers for anticipating and responding to disruption". This convergence between resilience and technological innovation has helped to enrich the concept, while raising new questions about governance and equity in smart, resilient cities.

Nevertheless, this development is not without its critics and debates. Some researchers, like Samuel Rufat (2015, p. 21), warn of the risks of "excessive and sometimes ill-considered use of the concept of resilience", stressing the need for a critical and contextualized approach to its application to urban studies. These debates reflect the inherent complexity of operationalizing such a rich, multi-dimensional concept in varied and constantly changing urban contexts.

In conclusion, the historical evolution of the application of the concept of resilience to urban studies bears witness to a dynamic process of adaptation and conceptual enrichment, reflecting the profound transformations of contemporary urban issues. This trajectory, marked by the shift from a vulnerability-centric approach to a more integrative and proactive vision of resilience, opens up new perspectives for rethinking urban development in the face of the challenges of the 21st century, while calling for continued vigilance as to the relevance and operationalization of the concept in diversified urban contexts.

### 1.1.3 Contextualizing urban resilience in the face of contemporary challenges

Contextualizing urban resilience in the face of contemporary challenges is part of a complex dynamic in which cities are at the heart of interconnected global and local issues, requiring a systemic and multidimensional approach. This contextualization reveals the urgent need to adopt resilience strategies adapted to a rapidly changing world, characterized by growing uncertainties and emerging risks.

Globalization and the growing interconnection of urban systems constitute a first major axis of this contextualization. As Pierre Veltz (2014, p. 23) points out, "cities have become the nodes of a global network of economic, cultural and informational exchanges, amplifying both their development opportunities and their vulnerability to systemic shocks". This increased interdependence of urban systems on a global scale redefines the contours of urban resilience, requiring consideration of cascading effects and complex feedbacks between local and global scales. Saskia Sassen's (2011, p. 117) work on "global cities" highlights this tension between the territorial anchorage of cities and their insertion in transnational networks, underlining the need for a multiscalar approach to urban resilience.

Climate change and increasing environmental risks constitute a second major challenge contextualizing contemporary urban resilience. According to the analyses of Magali Reghezza-Zitt and Samuel Rufat (2015, p. 45), "climate change acts as a threat multiplier, exacerbating existing vulnerabilities in urban systems and creating new long-term risks". This reality calls for a redefinition of urban resilience strategies, incorporating not only measures to adapt to climate impacts, but also far-reaching transformations of urban development models. The work of François Gemenne, Aleksandar Rankovic, Thomas Ansart, Benoît Martin, Patrice Mitrano and Antoine Rio (2019, p. 78) on the Atlas of the Anthropocene underlines the urgency of an ecological transition for cities, placing resilience at the heart of climate change mitigation and adaptation strategies.

The digital revolution and the emergence of "smart cities" constitute a third area of contextualization, redefining the ways in which urban resilience is understood and managed. As Antoine Picon (2018, p. 102) analyzes, "the massive integration of digital technologies into the urban fabric offers new opportunities to strengthen the resilience of cities, while raising crucial questions in terms of governance, data protection and social equity". This technological dimension of urban resilience calls for in-depth reflection on the link between technical and social innovation, in order to guarantee inclusive and democratic resilience.

The demographic and socio-economic transitions underway in many cities constitute a fourth major challenge contextualizing urban resilience. According to Laurent Davezies (2012, p. 89), "the dynamics of metropolization and urban shrinkage are redrawing the geography of vulnerability and resilience on a territorial scale". These mutations require resilience strategies to be adapted to the specificities of urban trajectories, taking into account issues of social cohesion, spatial justice and sustainable economic development.

Finally, the global health crisis linked to COVID-19 has brought into sharp focus the need for a holistic approach to urban resilience. As Michel Lussault and Johanna Lévy (2020, p. 31) point out, "the pandemic revealed the complex interconnections between public health, spatial organization, social dynamics and the economic resilience of cities". This crisis has accelerated awareness of the importance of integrating health and well-being issues into urban resilience strategies, beyond mere environmental or economic considerations.

In conclusion, the contextualization of urban resilience in the face of contemporary challenges reveals the complexity and interdependence of the issues facing 21st century cities. This contextualization underlines the need for a systemic and adaptive approach to resilience, capable of integrating the multiple and evolving dimensions of urban vulnerabilities and potentialities. It also calls for in-depth reflection on the forms of governance and citizen participation needed to build resilient, inclusive and sustainable cities in the face of the growing uncertainties of today's world.

## 1.1.4 Criticism and debate surrounding the adoption of the resilience paradigm

The resilience paradigm, while widely adopted in urban studies and planning policies, is the subject of significant criticism and debate within the scientific and practitioner community. These discussions raise important questions about the relevance, effectiveness and ethical implications of applying the concept of resilience in the urban context.

A first line of criticism concerns the conceptual ambiguity and malleability of the term "resilience", which can lead to divergent interpretations and a dilution of its operational potential. As Pierre-André Julien (2019, p. 78) points out, "the polysemy of the concept of urban resilience, while favoring its adoption by various actors, risks compromising its ability to effectively guide public action and urban planning". This criticism highlights the danger of a superficial or rhetorical use of resilience, without any real commitment to structural transformations of urban systems.

Another line of debate concerns the political and social implications of adopting the resilience paradigm. Some researchers, like Béatrice Quenault (2017, p. 145), argue that "the discourse on urban resilience can serve to legitimize a disengagement of the state and a transfer of responsibilities to local communities and individuals, without necessarily providing them with adequate resources". This critical perspective raises important questions about equity and social justice in the implementation of urban resilience strategies, particularly in contexts of high socio-economic inequality.

The tension between resilience and transformation is another major point of debate. As François Mancebo (2018, p. 212) explains, "the emphasis on resilience can sometimes lead to an excessive focus on maintaining existing structures, to the detriment of more radical transformations potentially necessary in the face of contemporary urban challenges". This criticism questions the ability of the resilience paradigm to promote profound systemic change, particularly in the context of the climate crisis and urgent socio-ecological transitions.

Finally, epistemological and methodological

questions are raised regarding the measurement and assessment of urban resilience. Magali Reghezza-Zitt and Samuel Rufat (2021, p. 89) point out that "the quantification of urban resilience, while attractive to decision-makers, faces considerable conceptual and practical challenges, running the risk of reducing the complexity of urban dynamics to simplistic indicators". This debate highlights the limits of purely quantitative approaches and calls for more nuanced and contextualized methodologies in the study of urban resilience.

Far from discrediting the concept of urban resilience, these criticisms and debates contribute to its enrichment and refinement. They call for in-depth reflection on how the resilience paradigm can be applied to urban policies, taking into account issues of equity, systemic transformation and the complexity of urban systems. As Cyria Emelianoff (2020, p. 167) concludes, "the future of urban resilience lies in its ability to integrate these critiques to evolve towards a more robust, equitable and transformative conceptual and operational framework".

## 1.2 Definitions and interpretations of urban resilience

### 1.2.1 Comparative analysis of existing definitions

A comparative analysis of existing definitions of urban resilience reveals a conceptual diversity that reflects the complexity and multidimensionality of this paradigm. This definitional variability, far from being a simple semantic exercise, has profound implications for the way resilience is conceptualized, measured and implemented in urban policies.

A frequently cited approach is that proposed by Serge Lhomme, Damien Serre, Richard Laganier and Youssef Diab (2013, p. 25), who define urban resilience as "the capacity of an urban system to absorb a disturbance and recover its functions following this disturbance". This definition emphasizes the notions of absorption and recovery, underlining the dynamic dimension of resilience. However, as the authors note, this perspective can be criticized for its potentially too narrow focus on the return to a pre-disturbance state, thus neglecting the transformative aspects of resilience.

From a more holistic perspective, Magali Reghezza- Zitt and Samuel Rufat (2015, p. 42) propose to conceive of urban resilience as "a dynamic process of positive adaptation enabling cities to cope, adapt and transform in the face of stresses and shocks, whether sudden or chronic". This definition broadens the scope of the concept by explicitly integrating the notions of adaptation and transformation, reflecting a more nuanced understanding of the possible trajectories of urban systems in the face of disruption.

Another influential approach is that developed by Béatrice Quenault (2017, p. 56), who defines urban resilience as "the ability of urban systems to maintain or rapidly recover desired functions in the face of disruption, to adapt to change and to rapidly transform systems that limit current or future adaptive capacity". This definition is particularly interesting as it explicitly incorporates the notion of "desired functions", thus introducing a normative dimension that underlines the importance of societal choices in determining resilience objectives.

On the practitioner and institutional side, the definition proposed by the Rockefeller Foundation as part of its "100 Resilient Cities" program, and adopted by many French cities, sees urban resilience as "the ability of individuals, communities, institutions, businesses and systems within a city to survive, adapt and thrive regardless of the types of chronic stresses and acute crises they experience" (quoted in Emeline Comby, 2020, p. 89). This definition is distinguished by its inclusive nature, encompassing various scales of action and types of urban actors.

The variability of these definitions reflects not only disciplinary and contextual differences, but also fundamental conceptual tensions. As François Mancebo (2018, p. 134) points out, "the multiplicity of definitions of urban resilience reflects a persistent tension between descriptive approaches, aiming to characterize the dynamics of urban systems, and normative approaches, seeking to prescribe objectives and means for strengthening resilience". This tension between description and prescription is at the heart of numerous debates on the operationalization of the concept in urban policies.

In conclusion, a comparative analysis of existing definitions of urban resilience reveals a rich and dynamic conceptual field. As Cyria Emelianoff (2021, p. 78) suggests, "rather than seeking a single, consensual definition, the challenge lies in the ability to articulate these different perspectives to develop a finer, operational understanding of urban resilience". This definitional diversity, while presenting challenges for research and practice, also offers the flexibility to adapt the concept to the specificities of different urban contexts and to the varied objectives of the actors involved in building more resilient cities.

## 1.2.2 Key components of urban resilience

By analyzing the key components of urban resilience, we can draw up a conceptual framework for understanding and operationalizing this complex concept. Although perspectives vary according to authors and disciplines, certain fundamental elements emerge repeatedly in the scientific literature.

The capacity to absorb shocks is a first essential component of urban resilience. As Serge Lhomme and Damien Serre (2015, p. 42) point out, "this capacity translates into the ability of the urban system to maintain its essential functions in the face of sudden and intense disturbances". This dimension of resilience emphasizes the robustness of urban infrastructures and the redundancy of critical systems, enabling the city to continue functioning even in a crisis situation.

Adaptability in the face of gradual change

represents a second key component, particularly relevant in the context of the chronic stresses faced by contemporary cities. Magali Reghezza- Zitt (2017, p. 89) defines this adaptability as "the ability of the urban system to adjust its structures and processes in response to gradual changes in its environment, without losing its fundamental identity". This component underlines the importance of flexibility and continuous learning in urban management.

Transformability and systemic innovation constitute a third essential component, particularly emphasized in the most recent approaches to urban resilience. François Mancebo (2018, p. 156) argues that "true urban resilience is not limited to shock absorption or incremental adaptation, but involves the ability to fundamentally rethink urban systems to meet emerging challenges". This transformative dimension of resilience opens the way to radical innovations in city design and management.

Beyond these three widely recognized components, other elements are frequently cited as constitutive of urban resilience. Cyria Emelianoff (2019, p. 112) stresses the importance of "connectivity and social networks as key factors of resilience, facilitating the flow of information and the mobilization of resources in times of crisis". This perspective highlights the social and relational dimension of urban resilience, beyond purely technical or infrastructural aspects.

The diversity and modularity of urban systems

are also considered essential components of resilience. As Bruno Barroca and Damien Serre (2016, p. 78) explain, "a diverse and modular urban system is less vulnerable to cascading failures and more able to adapt to changing conditions". This approach advocates decentralized, polycentric urban organization, promoting resilience at different scales.

Finally, the capacity for self-organization and collective learning is increasingly recognized as a crucial component of urban resilience. Samuel Rufat (2020, p. 203) argues that "urban resilience rests largely on the capacity of local communities to organize, innovate and learn from their experiences". This perspective emphasizes the importance of citizen participation and collaborative governance in building resilient cities.

Articulating these different components poses considerable conceptual and practical challenges. As Béatrice Quenault and Helga-Jane Scarwell (2021, p. 67) note, "urban resilience cannot be reduced to the sum of its components, but must be understood as a complex system of interactions between these different elements". This systemic approach underlines the need for a holistic and integrated vision in the planning and management of urban resilience.

In conclusion, the identification and analysis of key components of urban resilience provides a rich conceptual framework to guide research and action in this field. However, as Emeline Comby (2022, p. 145) reminds us, "the relevance and prioritization of these components may vary according to specific urban contexts, calling for a flexible and adaptive approach in their application". This observation calls for ongoing reflection on how these components articulate and express themselves in different urban contexts, thus contributing to the evolution and enrichment of the concept of urban resilience.

### 1.2.3 Specific vs. general resilience in the urban context

The distinction between specific resilience and general resilience in the urban context is a major line of thought, offering complementary perspectives on how cities can cope with the multiple and complex challenges they face.

Specific resilience, as defined by Magali Reghezza-Zitt and Samuel Rufat (2015, p. 98), "concerns the ability of an urban system to resist and recover from a particular type of disruption or stress". This approach targets identified threats and specific vulnerabilities, enabling targeted planning and preparation. For example, a coastal city can develop specific resilience to the risks of flooding or rising sea levels, by putting in place appropriate protective infrastructures and early warning systems.

Conversely, general resilience, according to François Mancebo (2018, p. 167), "refers to the overall capacity of an urban system to cope with unforeseen disturbances and adapt to multiple and sometimes unknown changes". This more holistic approach aims to strengthen the city's overall adaptive capacities, irrespective of the specific nature of the challenges ahead. It emphasizes the flexibility, resource diversity and learning capacity of the urban system as a whole.

The tension between these two approaches raises important questions for urban planning and governance. As Béatrice Quenault (2017, p. 134) points out, "the balance between specific and general resilience is a crucial issue for cities faced with limited resources and competing priorities". Indeed, investing in specific resilience can offer more immediate and measurable results in the face of known risks, but can potentially leave the city vulnerable to unforeseen threats.

Cyria Emelianoff (2020, p. 212) argues that "optimal urban resilience requires a judicious integration of specific and general approaches, tailored to the local context and the city's capabilities". This perspective underlines the importance of a strategic approach that combines targeted measures to address priority risks with investments in the city's general adaptive capacities.

The work of Serge Lhomme and Damien Serre (2019, p. 76) highlights the potential synergies between specific and general resilience: "Efforts to build resilience to a specific risk can often contribute to improving the overall resilience of the city, provided they are designed from a systemic perspective". For example, urban greening measures to combat heat islands can also improve overall quality of life and urban biodiversity, thereby strengthening the city's overall resilience.

However, Bruno Barroca and Gilles Hubert (2021, p. 189) warn against an excessive focus on specific resilience to the detriment of general resilience: "An approach that focuses too much on specific risks can lead to the neglect of systemic vulnerabilities and limit the city's ability to cope with emerging or unforeseen challenges". This perspective argues for a balanced approach that does not lose sight of the importance of the overall flexibility and adaptability of the urban system.

The question of spatial and temporal scale is also crucial to this discussion. As Helga-Jane Scarwell and Olivier Petit (2016, p. 145) note, "specific resilience may be more relevant at the local and short-term scale, while general resilience becomes more important at the metropolitan and long-term scale". This observation underlines the need for a multi-scale approach to urban resilience planning.

In conclusion, the debate between specific and general resilience in the urban context reflects the complexity of the challenges facing contemporary cities. As Samuel Rufat (2022, p. 234) summarizes, "the future of urban resilience lies in the ability to intelligently articulate these two approaches, taking into account local specificities, available resources and the uncertainties inherent in complex urban systems". This perspective calls for ongoing reflection on how to integrate these two dimensions of resilience into urban policies and practices, in order to build cities capable of facing the known and unknown challenges of the 21st century.

## 1.2.4 Interactions between urban resilience and related concepts

The study of urban resilience cannot be carried out in isolation, as this concept is part of a complex network of interactions with other key notions of contemporary urbanism. By analyzing these interactions, we can not only refine our understanding of urban resilience, but also better grasp its place in the broader panorama of current urban issues.

The interplay between urban resilience and urban sustainability is particularly significant. As Cyria Emelianoff (2019, p. 78) points out, "urban resilience and sustainability, while distinct, share common goals and are mutually reinforcing". Sustainability aims to balance present and future needs, while resilience focuses on the ability to cope with disruption. However, these two concepts converge in their ambition to create more adaptable and perennial cities. François Mancebo (2018, p. 123) argues that "a truly sustainable city must necessarily be resilient, and vice versa", underlining the intrinsic complementarity of these two approaches.

The relationship between urban resilience and vulnerability is also crucial. Magali Reghezza-Zitt and

Samuel Rufat (2015, p. 156) explain that "resilience cannot be understood without taking into account the specific vulnerabilities of urban systems". Indeed, urban resilience often aims to reduce these vulnerabilities, but can sometimes mask or displace them. Béatrice Quenault (2017, p. 89) warns against "an approach to resilience that neglects in-depth analysis of structural vulnerabilities", stressing the importance of an integrated vision of these two concepts.

The interaction between urban resilience and risk management is also significant. Serge Lhomme and Damien Serre (2020, p. 112) argue that "urban resilience offers a broader and more dynamic framework than traditional risk management, integrating not only prevention and preparedness, but also adaptation and long-term transformation". This broader perspective makes it possible to approach urban risks in a more holistic and proactive way.

The concept of "smart cities" also has close links with urban resilience. As Bruno Barroca and Gilles Hubert (2021, p. 178) explain, "smart technologies can strengthen urban resilience by improving data collection and analysis, real-time communication and adaptive management of urban resources". However, they also warn against over-reliance on technology, which could create new vulnerabilities.

Urban innovation, in its broadest sense, is intrinsically linked to resilience. Helga-Jane Scarwell and Olivier Petit (2018, p. 201) point out that "a city's capacity for innovation is a key factor in its resilience, enabling it to adapt creatively to emerging challenges". This perspective highlights the importance of experimentation and continuous learning in building resilient cities.

Spatial and social justice is another concept closely linked to urban resilience. Emeline Comby (2022, p. 167) argues that "urban resilience that does not take justice issues into account risks reinforcing existing inequalities". This reflection underlines the importance of an inclusive and equitable approach when implementing resilience strategies.

Finally, the concept of urban ecology offers enriching perspectives for urban resilience. Samuel Rufat (2021, p. 245) explains that "the ecosystem approach to the city, inspired by urban ecology, makes it possible to conceptualize urban resilience as an emergent property of a complex, adaptive system". This systemic vision enriches our understanding of resilience dynamics on an urban scale.

In conclusion, as Cyria Emelianoff (2020, p. 289) summarizes, "urban resilience cannot be apprehended in isolation, but must be understood as an integral part of a broader conceptual ecosystem". This perspective calls for a transdisciplinary and integrated approach to urban resilience, taking into account its multiple interactions with other key concepts of contemporary urbanism. Such an approach not only enriches our theoretical understanding of urban resilience, but also informs more coherent and effective urban policies and practices in the face of the complex challenges of the 21st century.

## 1.3 Urban resilience as a dynamic, multiscalar process

## 1.3.1 Temporalities of urban resilience

Analyzing the temporalities of urban resilience provides an essential framework for understanding the complex dynamics governing cities' ability to cope with disruption and adapt to change. This temporal approach makes it possible to distinguish different phases of resilience, each presenting specific challenges and strategies.

> ➢ *Short-term response to acute shocks*

In the immediate term, urban resilience manifests itself in the city's ability to respond effectively to acute shocks. As Magali Reghezza-Zitt and Samuel Rufat (2015, p. 123) point out, "short-term resilience focuses on crisis management and the ability of the urban system to maintain its essential functions in the face of sudden disruption".

This phase involves the rapid mobilization of resources, the coordination of players and the activation of emergency plans.

Serge Lhomme and Damien Serre (2018, p. 89) point out that "short-term resilience relies heavily on the robustness of critical infrastructures and the effectiveness of crisis management systems". They stress the importance of system redundancy and operational flexibility to ensure a rapid and effective response to disruptions.

## ➤ *The medium term: adapting to chronic stress*

On a broader time scale, urban resilience is the ability to adapt to chronic stresses. François Mancebo (2020, p. 156) defines this phase as "a continuous process of adjustment and reconfiguration of urban systems in the face of persistent pressures". The aim here is to respond to challenges such as climate change, demographic transitions and economic mutations.

Béatrice Quenault (2017, p. 201) points out that "medium-term adaptation requires a more strategic and anticipatory approach, involving gradual changes to urban policies, planning practices and social behaviors". This perspective highlights the importance of organizational learning and incremental innovation in building urban resilience.

## ➤ *The long term: transformation and evolution of urban systems*

In the long term, urban resilience implies deeper transformations of urban systems. Cyria Emelianoff (2021, p. 278) argues that "long-term resilience requires a capacity for radical transformation, enabling cities to reinvent themselves in the face of major structural change". This transformative dimension of resilience opens the way to disruptive innovations and fundamental reconfigurations of urban organization.

Bruno Barroca and Gilles Hubert (2019, p. 167) point out that "long-term resilience implies a forward-looking vision and an ability to anticipate and shape possible urban futures". This approach emphasizes the importance of long-term strategic planning and scenario-building to guide urban development trajectories.

> ➢ *Interactions and temporal overlaps*

Crucially, these different temporalities of urban resilience are not mutually exclusive, but interact in complex ways. As Helga-Jane Scarwell and Olivier Petit (2020, p. 234) explain, "actions taken in the short term in response to a crisis can have long-term implications for the city's capacity for transformation, while long-term strategies influence the ability to respond immediately to shocks".

Samuel Rufat (2022, p. 312) stresses the importance of a "multi-temporal approach to urban resilience, capable of articulating immediate responses, gradual adaptations and long-term transformations". This perspective calls for an integrated and dynamic management of resilience, taking into account different temporal scales and their interactions.

In conclusion, as Emeline Comby (2021, p. 189) summarizes, "urban resilience unfolds along a temporal continuum, from immediate response to crises to long-term transformation of urban systems". This enriched temporal vision of urban resilience offers a powerful conceptual framework to guide urban planning and management in the face of the multiple and complex challenges of the 21st century. It underlines the need for a flexible and adaptive approach, capable of navigating the different temporalities of resilience to build more robust, adaptable and transformative cities.

## 1.3.2 Spatial scales of urban resilience

Analyzing the spatial scales of urban resilience offers an essential perspective for understanding the complexity and interconnectedness of urban systems. This multi-scalar approach makes it possible to understand resilience at different levels, from local to global, and to examine the interactions between these scales.

> ➢ *From neighborhood to metropolis: articulating intra-urban scales*

At the finest scale, urban resilience manifests itself at the neighborhood level. As Magali Reghezza-Zitt and Samuel Rufat (2017, p. 156) point out, "the neighborhood is often the basic unit of urban resilience, where social ties are forged and local initiatives develop". This scale makes it possible to observe and strengthen communities' capacity for self-organization in the face of everyday challenges and disruptions.

However, resilience at the neighborhood level cannot be isolated from broader city dynamics. François Mancebo (2019, p. 234) argues that "urban resilience requires a coherent articulation between neighborhood initiatives and city-wide policies". This perspective underlines the importance of multi-level governance capable of integrating local actions into a global urban strategy.

On a metropolitan scale, resilience takes on a more systemic dimension. Serge Lhomme and Damien Serre (2020, p. 178) explain that "metropolitan resilience involves the management of complex, interdependent urban systems, requiring large-scale coordination". This scale makes it possible to address issues such as critical infrastructure management, urban mobility and long-term strategic planning.

> ***Urban resilience in a regional and national context***

Beyond the metropolis, urban resilience is part of a broader regional and national context. Béatrice Quenault (2018, p. 289) points out that "a city's resilience cannot be dissociated from its regional environment, particularly in terms of resources, economic flows and risk management". This perspective highlights the importance of city-countryside relations and territorial dynamics in building urban resilience.

At a national level, policies and regulatory frameworks play a crucial role in defining and implementing urban resilience. Cyria Emelianoff (2021, p. 312) argues that "national resilience strategies strongly influence cities' resilience capacities and orientations". This scale makes it possible to address issues such as national infrastructure planning, decentralization policies or climate change adaptation strategies.


> ### ➢ *Global connections and the resilience of city networks*

On a global scale, urban resilience is taking on a new dimension, linked to the growing interconnection of cities in a globalized world. Bruno Barroca and Gilles Hubert (2019, p. 201) explain that "the resilience of cities is increasingly dependent on their ability to navigate global networks of economic, cultural and knowledge exchange". This perspective underlines the importance of international partnerships and exchanges of best practice between cities around the world.

Helga-Jane Scarwell and Olivier Petit (2020, p. 145) put forward the concept of "networked resilience", arguing that "the resilience of a city cannot be understood in isolation, but must be analyzed in the context of the global urban networks to which it belongs". This approach invites us to consider urban resilience as an emerging property of globally interconnected city systems.

In conclusion, as Magali Reghezza-Zitt (2023, p. 334) summarizes, "the multi-scalar approach to urban resilience offers a powerful conceptual framework for understanding the complexity of contemporary urban challenges". This enriched vision of urban resilience underlines the importance of integrated planning and management, capable of navigating between different spatial scales to build more resilient cities and urban systems in the face of the challenges of the 21st century.

It calls for a rethinking of urban governance methods to take better account of the complex interactions between the local and the global in building resilience.

## 1.3.3 Interactions between scales and threshold effects

Analyzing interactions between scales and threshold effects in the context of urban resilience offers a crucial perspective for understanding the complexity of urban systems and the dynamics that govern their ability to cope with disturbance and adapt to change.

François Mancebo (2018, p. 178) points out that "interactions between the different spatial and temporal scales of urban resilience are not linear, but characterized by threshold effects and complex feedback loops". This observation highlights the systemic nature of urban resilience, where minor changes at one scale can potentially trigger major transformations at other scales.

Threshold effects, in particular, play a crucial role in the dynamics of urban resilience. As Magali Reghezza-Zitt and Samuel Rufat (2019, p. 234) explain, "urban systems can absorb disturbances up to a certain point, beyond which abrupt and sometimes irreversible changes can occur". These threshold effects can manifest themselves at different scales, from the neighborhood to the metropolis, and in various domains, whether physical infrastructures, social systems or ecological processes.

Serge Lhomme and Damien Serre (2020, p. 156) put forward the concept of "cascades of effects" in their analysis of interactions between scales. They argue that "a disturbance at one scale can propagate through the different levels of the urban system, creating cascading effects that amplify the initial impact". This perspective underlines the importance of a holistic approach to resilience, capable of taking account of these complex interdependencies.

Interactions between scales can also create positive synergies for urban resilience. Béatrice Quenault (2021, p. 289) observes that "successful local initiatives can be amplified and disseminated on a city-wide scale, and even beyond, creating a positive ripple effect for global resilience". This bottom-up dynamic underlines the potential of local innovations in building resilience on a larger scale.

However, these interactions can also reveal tensions and contradictions between the different scales of resilience. Cyria Emelianoff (2022, p. 201) notes that "measures aimed at strengthening resilience at one scale can sometimes compromise resilience at another". For example, urban densification policies to improve energy efficiency can, in some cases, reduce local adaptive capacity in the face of heat islands.

Threshold effects and interactions between scales also raise important questions in terms of governance and decision-making. Bruno Barroca and Gilles Hubert (2023, p. 178) argue that "effective management of urban resilience requires adaptive governance mechanisms capable of detecting and responding to threshold effects at different scales". This approach implies the development of sophisticated monitoring systems and flexible decision-making processes.

The temporal dimension adds a further layer of complexity to these interactions. Helga-Jane Scarwell and Olivier Petit (2020, p. 312) point out that "threshold effects and interactions between scales can manifest themselves over highly variable temporalities, ranging from the immediate to the long term". This observation calls for a forward-looking approach to urban resilience, capable of anticipating and managing these complex dynamics over different time scales.

The analysis of interactions between scales and threshold effects also reveals the importance of transdisciplinary approaches in the study of urban resilience. Samuel Rufat (2021, p. 245) argues that "understanding these complex dynamics requires the integration of knowledge from multiple disciplines, from urban ecology and sociology to engineering and economics". This perspective underlines the need for greater collaboration between researchers, practitioners and decision-makers to understand urban resilience in all its complexity.

In conclusion, as Emeline Comby (2023, p. 189) summarizes, "the study of interactions between scales and threshold effects in urban resilience offers a powerful conceptual framework for rethinking the planning and management of cities in the face of the challenges of the 21st century". This approach calls for the development of more nuanced and adaptive resilience strategies, capable of navigating the complexity of urban systems and taking advantage of synergies between different scales while minimizing the risk of negative cascades. It also underlines the importance of a systemic and integrated vision of urban resilience, moving beyond traditional sectoral approaches to embrace the interconnected and dynamic nature of contemporary cities.

## 1.3.4 Multi-level governance and stakeholder coordination

Multi-level governance and stakeholder coordination are fundamental to the effective implementation of urban resilience. This approach recognizes the complexity of urban systems and the need for concerted action across different scales and sectors.

François Mancebo (2019, p. 201) points out that "urban resilience requires adaptive governance capable of articulating actions at different scales, from local to global, and coordinating a multiplicity of actors with sometimes divergent interests". This perspective highlights the challenge of creating flexible, inclusive governance structures capable of adapting to the changing dynamics of urban systems.

Vertical coordination between different levels of government plays a crucial role in the governance of urban resilience. As Magali Reghezza-Zitt and Samuel Rufat (2020, p. 156) explain, "effective articulation between national, regional and local policies is essential to ensure a coherent and integrated approach to resilience". This vertical coordination aligns long-term strategic objectives with concrete actions on the ground.

At the same time, horizontal coordination between different sectors and actors within the same scale is equally important. Béatrice Quenault (2021, p. 278) argues that "urban resilience requires a cross-sectoral approach, overcoming traditional sectoral silos to foster integrated action". This perspective underlines the importance of creating platforms for cross-sector collaboration, bringing together actors from the public, private and civil society sectors.

Citizen participation is emerging as a key element in the multi-level governance of urban resilience. Cyria Emelianoff (2022, p. 189) observes that "the active involvement of citizens in resilience planning and implementation processes strengthens the legitimacy and effectiveness of the actions undertaken". This participatory approach mobilizes local knowledge and promotes community ownership of resilience strategies.

Knowledge management and information sharing between the different levels and players is another major challenge. Serge Lhomme and Damien Serre (2020, p. 234) point out that "the effective circulation of information and knowledge between the different levels of governance is crucial for informed decision-making on resilience". This implies the development of integrated information systems and mechanisms for sharing best practices.

The question of resources and capacities at different levels of governance is also central. Bruno Barroca and Gilles Hubert (2023, p. 167) note that "the equitable distribution of resources and capacity building at all levels are essential for effective governance of urban resilience". This observation underlines the importance of innovative financing mechanisms and capacity-building programs.

City networks and international partnerships are playing a growing role in the multi-level governance of urban resilience. Helga-Jane Scarwell and Olivier Petit (2021, p. 312) argue that "exchanges of experience and cooperation between cities on an international scale offer valuable opportunities to strengthen local resilience". These networks make it possible to pool resources, share knowledge and influence political agendas at different scales.

The temporality of governance is another crucial aspect. Samuel Rufat (2022, p. 245) points out that "the governance of urban resilience must be able to reconcile short-term needs with long-term strategic visions". This approach implies the development of adaptive planning mechanisms, capable of adjusting strategies as contexts and knowledge evolve.

Tensions and conflicts between different actors and scales of governance represent an ongoing challenge. Emeline Comby (2023, p. 178) observes that "the multi-level governance of urban resilience often involves complex negotiations and compromises between divergent interests". This perspective underlines the importance of mediation and conflict resolution skills in urban resilience management.

In conclusion, as summarized by Magali Reghezza-Zitt (2020, p. 301), "multi-level governance and stakeholder coordination are the keystones of an integrated and effective approach to urban resilience". This vision of governance calls for a rethinking of institutional structures and decision-making processes to better reflect the complex and dynamic nature of urban systems. It also underlines the importance of a culture of collaboration and continuous learning, capable of navigating uncertainty and adapting to the emerging challenges of the 21st century. The governance of urban resilience thus appears to be an iterative and collaborative process, requiring the sustained and coordinated involvement of all stakeholders at every scale of the city.

## 1.4 Dimensions of urban resilience: economic, social, environmental and institutional

### 1.4.1 Economic resilience

Economic resilience is a fundamental dimension of overall urban resilience, reflecting a city's ability to maintain its economic vitality in the face of chronic shocks and stresses. This component of urban resilience encompasses several interconnected aspects that merit in-depth analysis.

## ➢ *Economic diversification and innovation*

Diversification of the economic fabric is emerging as a central pillar of urban economic resilience. As François Mancebo and Cyria Emelianoff (2018, p. 156) point out, "a diversified urban economy is less vulnerable to sectoral shocks and better able to adapt to structural changes". This approach implies cultivating a balance between different economic sectors, ranging from traditional industries to innovative services.

Innovation plays a crucial role in this diversification dynamic. Béatrice Quenault (2020, p. 189) argues that "a city's ability to stimulate innovation, particularly through dynamic entrepreneurial ecosystems, strengthens its long-term economic resilience". This perspective highlights the importance of innovation support policies, business incubators and partnerships between research and industry.

## ➢ *Labor markets and human capital*

The resilience of urban labor markets is another key aspect of economic resilience. Magali Reghezza-Zitt and Samuel Rufat (2021, p. 234) observe that "the flexibility and adaptability of the urban workforce are essential to absorb economic shocks and seize new opportunities". This approach underlines the importance of continuous training and retraining policies to maintain the employability of the workforce.

Developing human capital appears to be a strategic investment for economic resilience. Serge Lhomme and Damien Serre (2019, p. 278) argue that "the education and training of the urban population is a key resilience factor, enabling better adaptation to economic change". This perspective highlights the importance of investment in education, research and skills development.

➤ *Financial and budgetary resilience*

The financial health of local authorities plays a crucial role in urban economic resilience. Bruno Barroca and Gilles Hubert (2022, p. 201) point out that "the ability of cities to maintain fiscal stability and access diversified financing strengthens their resilience to economic shocks". This observation highlights the importance of prudent financial management and strategies to diversify municipal revenue sources.

The question of debt and financial risk management is also central. Helga-Jane Scarwell and Olivier Petit (2020, p. 312) note that "the financial resilience of cities implies proactive risk management and long-term investment planning". This approach underlines the importance of robust and transparent financial governance mechanisms.

➤ *Circular economy and localization*

The emergence of the circular economy offers new perspectives for urban economic resilience.

Emeline Comby (2023, p. 167) argues that "adopting circular economy principles can strengthen local economic resilience by reducing dependence on external resources and creating new economic opportunities". This approach emphasizes the importance of policies that encourage recycling, reuse and valorization of local resources.

The relocation of certain economic activities is also emerging as a resilience strategy. Samuel Rufat (2021, p. 245) observes that "the COVID-19 pandemic has highlighted the importance of shorter supply chains and a degree of local economic autonomy for urban resilience". This perspective invites us to rethink the balance between globalization and the local anchoring of economic activities.

In conclusion, as Cyria Emelianoff (2020, p. 289) summarizes, "urban economic resilience appears to be a multidimensional process, requiring an integrated approach that takes into account economic diversification, human capital development, the financial health of communities and innovation in economic models". This vision of economic resilience calls for a rethinking of urban development strategies to better integrate the principles of flexibility, adaptability and sustainability. It also underlines the importance of collaborative economic governance, involving a diversity of public, private and civil society players in building more resilient urban economies to meet the challenges of the 21st century.

## 1.4.2 Social resilience

Social resilience is a crucial dimension of overall urban resilience, reflecting the ability of urban communities to adapt, transform and thrive in the face of challenge and disruption. This component of urban resilience encompasses several interconnected aspects that merit in-depth analysis.

> ### *Social cohesion and social capital*

Social cohesion is emerging as a fundamental pillar of urban social resilience. As Magali Reghezza-Zitt and Samuel Rufat (2019, p.

178), "strong social cohesion enables urban communities to cope better with crises by mobilizing collective resources and fostering mutual support". This observation highlights the importance of social ties, mutual trust and a sense of belonging in building resilient cities.

The concept of social capital plays a central role in this dynamic. François Mancebo (2020, p. 234) argues that "social capital, understood as the set of networks and norms of reciprocity within a community, constitutes an essential resource for urban resilience". This perspective underlines the importance of initiatives promoting social interaction, volunteering and civic engagement.

> ### *Equity and inclusion in access to urban resources*

Equity of access to urban resources and services

appears to be a key element of social resilience. Béatrice Quenault (2021, p. 156) observes that "a socially resilient city is one that ensures equitable access to its resources, thereby reducing differential vulnerability to crises". This approach highlights the importance of inclusive urban policies, particularly in terms of housing, health, education and mobility.

Combating socio-spatial inequalities is part of this approach. Cyria Emelianoff (2022, p. 289) points out that "reducing disparities between neighborhoods and promoting social diversity strengthens the city's overall resilience by reducing points of fragility". This observation invites us to rethink urban planning from the perspective of spatial justice.

## ➤ *Citizen participation and community empowerment*

The active participation of citizens in urban governance is emerging as a crucial factor in social resilience. Serge Lhomme and Damien Serre (2020, p. 201) argue that "involving residents in decision-making and urban planning processes strengthens their capacity for collective action in the face of challenges". This perspective highlights the importance of participatory democracy mechanisms and the co-construction of urban policies.

The empowerment of local communities plays a central role in this dynamic. Helga-Jane Scarwell and Olivier Petit (2023, p. 312) note that "strengthening communities' capacity for action and self-organization is a powerful lever for urban social resilience". This approach underlines the importance of community development initiatives and the social economy.

## ➢ *Cultural diversity and resilience*

Cultural diversity appears as a resource for urban social resilience. Bruno Barroca and Gilles Hubert (2021, p. 167) observe that "a city's cultural diversity can be a source of creativity and innovation in problem-solving, strengthening its capacity to adapt". This perspective invites us to value diversity as an asset for resilience, while taking care to promote intercultural dialogue and inclusion.

## ➢ *Health and well-being*

The health and well-being of urban populations is a fundamental aspect of social resilience. Samuel Rufat (2022, p. 245) points out that "a physically and mentally healthy population is better able to cope with urban shocks and stresses". This observation highlights the importance of public health policies, promoting well-being and creating healthy urban environments.

## ➢ *Education and lifelong learning*

Education emerges as a crucial lever for urban social resilience. Emeline Comby (2021, p. 189) argues that "access to quality education and lifelong learning opportunities strengthens the adaptive capacity of individuals and communities in the face of change". This perspective underlines the importance of investment in education systems and lifelong learning.

In conclusion, as summarized by Magali Reghezza-Zitt (2023, p. 301), "urban social resilience appears as a multidimensional process, requiring an integrated approach that takes into account social cohesion, equity, citizen participation, cultural diversity, health and education". This vision of social resilience calls for a rethinking of urban policies from a more inclusive and participatory perspective, recognizing the central role of communities in building resilient cities. It also underlines the importance of a holistic approach to urban development, which considers social well-being and quality of life as essential components of urban resilience in the face of the challenges of the 21st century.

### 1.4.3 Environmental resilience

Environmental resilience is a fundamental dimension of overall urban resilience, reflecting the ability of urban systems to maintain their ecological functions and adapt to environmental change. This component encompasses several interconnected aspects that merit in-depth analysis.

➢ *Sustainable management of natural resources*

Sustainable natural resource management is emerging as a central pillar of urban environmental resilience. As Cyria Emelianoff and François Mancebo (2018, p. 156) point out, "an integrated approach to the management of water, energy and raw material resources is essential to ensure the long-term sustainability of urban systems". This observation highlights the importance of conservation, recycling and resource-use efficiency policies.

The circular economy plays a crucial role in this dynamic. Béatrice Quenault (2020, p. 189) argues that "the adoption of circular economy principles on an urban scale can significantly strengthen environmental resilience by reducing pressure on ecosystems and valorizing waste as a resource". This perspective underlines the importance of technological and organizational innovations in the management of urban material and energy flows.

## ➤ *Adapting to climate change*

Adapting to climate change is a major challenge for the environmental resilience of cities. Magali Reghezza-Zitt and Samuel Rufat (2021, p. 234) observe that "the ability of cities to anticipate and adapt to the impacts of climate change is crucial to their long-term resilience". This approach involves a variety of strategies, from adaptive urban planning to the implementation of green and blue infrastructure.

Reducing urban heat islands is emerging as a priority issue. Serge Lhomme and Damien Serre (2019, p. 278) point out that "urban greening and the management of impermeable surfaces play a key role in the thermal regulation of cities and their adaptation to global warming". This perspective highlights the importance of integrating nature-based solutions into urban planning.

## ➤ *Urban biodiversity and ecosystem services*

Preserving and restoring urban biodiversity are essential elements of environmental resilience. Bruno Barroca and Gilles Hubert (2022, p. 201) argue that "rich, functional urban biodiversity strengthens the capacity of urban ecosystems to resist disturbance and provide essential ecosystem services". This observation underlines the importance of ecological corridors, multifunctional green spaces and ecological management of urban spaces.

Urban ecosystem services play a central role in this dynamic. Helga-Jane Scarwell and Olivier Petit (2020, p. 312) note that "valuing urban ecosystem services, such as local climate regulation, air and water purification, and pollination, contributes significantly to the environmental resilience of cities". This approach calls for a rethinking of urban planning to better integrate and enhance these natural services.

➤ ***Environmental quality and public health***

Urban environmental quality is emerging as a crucial issue at the interface between environmental resilience and public health. Emeline Comby (2023, p. 167) points out that "reducing urban pollution and improving air, water and soil quality are essential to strengthening the health resilience of urban populations". This perspective highlights the importance of policies to combat pollution and promote a healthy urban environment.

➤ ***Energy transition and sustainable mobility***

Energy transition appears to be a major lever for

urban environmental resilience. Samuel Rufat (2024, p. 245) observes that "the development of renewable energies and the improvement of energy efficiency on an urban scale strengthen the autonomy and resilience of cities in the face of energy shocks". This approach underlines the importance of investment in sustainable energy infrastructure and smart grids.

Sustainable mobility also plays a crucial role. Cyria Emelianoff (2024, p. 289) argues that "the transition to modes of transport with low environmental impact contributes significantly to urban resilience by reducing greenhouse gas emissions and improving the quality of urban life". This perspective calls for a rethinking of urban planning and mobility systems to favor soft mobility and public transport.

## 1.4.4 Corporate resilience

Institutional resilience is an essential dimension of overall urban resilience, reflecting the ability of governance structures and organizations to adapt, innovate and respond effectively to urban challenges. This component encompasses several interconnected aspects that merit in-depth analysis.

### ➤ *Adaptive capacity of local institutions*

The adaptive capacities of local institutions are emerging as a central pillar of institutional resilience. As François Mancebo and Cyria Emelianoff (2019, p. 178) point out, "the flexibility and responsiveness of local administrative structures are crucial to responding effectively to crises and rapid changes in the urban context". This observation highlights the importance of ongoing training for public officials and the adoption of more agile management modes.

Organizational learning plays a crucial role in this dynamic. Béatrice Quenault (2021, p. 234) argues that "the ability of institutions to learn from past experience and integrate new knowledge significantly strengthens their resilience in the face of emerging challenges". This perspective underlines the importance of feedback and knowledge management mechanisms within public organizations.

> ➢ *Regulatory and policy frameworks promoting resilience*

The adaptation of regulatory and policy frameworks appears to be a key element of institutional resilience. Magali Reghezza-Zitt and Samuel Rufat (2020, p. 156) observe that "flexible and evolving regulatory frameworks are essential to enable innovation and rapid adaptation of urban policies in the face of change". This approach implies a regular review of standards and procedures to ensure their relevance to contemporary issues.

Long-term strategic planning is emerging as an important tool. Serge Lhomme and Damien Serre (2022, p. 289) point out that "the systematic integration of resilience principles into urban planning documents strengthens the capacity of institutions to anticipate and manage long-term risks". This perspective highlights the importance of a forward-looking vision in urban governance.

### ➢ *Public-private partnerships and cross-sector cooperation*

Collaboration between public and private players appears to be a major lever for institutional resilience. Bruno Barroca and Gilles Hubert (2023, p. 201) argue that "innovative public-private partnerships can bring complementary resources and skills, essential to meeting the complex challenges of urban resilience". This observation underlines the importance of flexible and transparent cooperation frameworks.

Cross-sectoral cooperation also plays a crucial role. Helga-Jane Scarwell and Olivier Petit (2021, p. 312) note that "the decompartmentalization of administrative services and the promotion of cross-sectoral approaches strengthen the capacity of institutions to address resilience issues holistically". This approach calls for a rethinking of organizational structures to foster collaboration and innovation.

### ➢ *Participatory governance and citizen inclusion*

The inclusion of citizens in governance processes is emerging as an essential factor in institutional resilience. Emeline Comby (2024, p. 167) points out that "the active involvement of citizens in decision-making and the implementation of resilience policies strengthens the legitimacy and effectiveness of institutional actions". This perspective highlights the importance of participatory democracy mechanisms and the co-construction of public policies.

## ➢ *Transparency and accountability*

Transparency and accountability appear to be pillars of institutional trust, essential to resilience. Samuel Rufat (2025, p. 245) observes that "transparent and accountable institutions are better able to mobilize public support and resources to deal with crises". This approach underlines the importance of democratic control mechanisms and open communication with citizens.

## ➢ *Institutional innovation and experimentation*

Institutional innovation is emerging as a crucial lever for resilience. Cyria Emelianoff (2020, p. 189) argues that "the capacity of institutions to experiment with new forms of organization and governance is essential to adapt to the emerging challenges of urban resilience". This perspective calls for the creation of spaces for experimentation and innovation within institutional structures.

In conclusion, as summarized by Magali Reghezza-Zitt (2022, p. 301), "urban institutional resilience appears as a multidimensional process, requiring an integrated approach that takes into account adaptive capacities, regulatory frameworks, partnerships, participatory governance, transparency and innovation". This vision of institutional resilience calls for an in-depth rethinking of urban governance modes to make them more flexible, inclusive and innovative. It also underlines the importance of an institutional culture focused on continuous learning and adaptation, capable of navigating the complexity and uncertainty of the urban challenges of the 21st century.

## 1.4.5 Interactions and synergies between the dimensions of urban resilience

The economic, social, environmental and institutional dimensions of urban resilience cannot be considered in isolation, but must be understood in their complex interactions and potential synergies. This systems approach is essential to understanding and strengthening the overall resilience of urban systems in the face of the multifaceted challenges they face (Quenault Béatrice, 2014, p. 89).

The interaction between economic and social resilience is particularly significant in the urban context. A diversified and innovative urban economy can foster job creation and social inclusion, strengthening the city's cohesion and social capital. Conversely, strong social cohesion and developed human capital can stimulate innovation and economic productivity (Toubin Marie, Lhomme Serge, Diab Youssef, Serre Damien & Laganier Richard, 2012, p. 631). This positive synergy between the economic and social dimensions can create a virtuous circle of urban resilience, where each dimension reinforces the other.

The environmental dimension of urban resilience interacts significantly with other dimensions. For example, the implementation of green infrastructure and nature-based solutions can not only improve environmental resilience to climate change, but also generate economic and social co-benefits. These infrastructures can create green jobs, improve residents' quality of life and enhance the attractiveness of the city (Lhomme Serge, Serre Damien, Diab Youssef & Laganier Richard, 2013, p. 15). In addition, sustainable natural resource management can contribute to long-term economic resilience by ensuring the continued availability of essential resources.

The institutional dimension plays a crucial cross-cutting role in urban resilience, facilitating and coordinating actions in the other dimensions. Adaptive local institutions and resilience-friendly regulatory frameworks can create an environment conducive to economic innovation, social cohesion and environmental sustainability (Rebotier Julien, 2012, p. 392). For example, integrated urban policies can simultaneously promote the functional mix of neighborhoods (economic dimension), social inclusion (social dimension) and energy efficiency (environmental dimension).

However, it is important to note that these interactions are not always synergistic, and can sometimes be contradictory. For example, some environmental resilience measures may conflict with short-term economic growth objectives. Similarly, the quest for greater economic efficiency can sometimes be at the expense of social cohesion or environmental sustainability (Rufat Samuel, 2012, p. 204). These tensions underline the importance of a balanced, integrated approach to urban resilience, which takes into account the potential trade-offs between different dimensions.

In conclusion, understanding and managing the interactions and synergies between the economic, social, environmental and institutional dimensions of urban resilience is essential to developing effective and sustainable resilience strategies. This holistic approach requires cross-sectoral collaboration, multi-level governance and a long-term vision of urban development (Toubin Marie, Lhomme Serge, Diab Youssef, Serre Damien & Laganier Richard, 2012, p. 640). It also offers opportunities to maximize co-benefits and create innovative solutions that simultaneously strengthen several aspects of urban resilience.

## 1.4.6 Challenges in measuring and assessing multidimensional urban resilience

Assessing and measuring multidimensional urban resilience presents considerable conceptual and methodological challenges. These difficulties are intrinsically linked to the complex and dynamic nature of urban systems, as well as to the multidimensionality of the concept of resilience itself.

One of the first challenges lies in defining and operationalizing the concept of urban resilience. As Quenault Béatrice, Pigeon Patrick, Bertrand Frédéric and Blond Nadège (2011, p. 23) point out, there is no clear consensus on precisely what resilience means in the urban context, which complicates the development of universally accepted indicators and metrics. The diversity of interpretations and disciplinary approaches to urban resilience makes it difficult to compare different studies and assessments.

The multidimensional nature of urban resilience also poses challenges in terms of integrating and weighting the various dimensions. How can we balance and combine economic, social, environmental and institutional indicators to obtain a coherent overall assessment of resilience? This question is all the more complex as the interactions between these dimensions are often non-linear and can vary according to specific urban contexts (Toubin Marie, Lhomme Serge, Diab Youssef, Serre Damien & Laganier Richard, 2012, p. 635).

Another major challenge concerns the temporality of urban resilience. Resilience is a dynamic process that manifests itself on different temporal scales, from the short term (immediate response to shocks) to the long term (transformation and adaptation). As Rufat Samuel (2012, p. 208) explains, measuring resilience at a given point in time may not adequately reflect a city's actual capacity to cope with future disruptions or adapt to gradual changes.

The question of spatial scale also poses difficulties. Urban resilience can be assessed at different levels, from the neighborhood to the metropolis, via regional and national scales. Each scale may require specific indicators and assessment methods, while having to take into account the interdependencies between these scales (Lhomme Serge, Serre Damien, Diab Youssef & Laganier Richard, 2013, p. 17).

Data availability and quality represent a major practical challenge. Many aspects of urban resilience, particularly in the social and institutional dimensions, are difficult to quantify and often require qualitative data. Moreover, the collection of consistent and internationally comparable data remains problematic, limiting the possibility of robust comparative analyses between different cities (Rebotier Julien, 2012, p. 395).

Finally, there is debate about the very relevance of seeking to measure urban resilience quantitatively. Some researchers argue that the complexity and contextual specificity of urban resilience cannot be fully captured by quantitative indicators, and argue for more qualitative and participatory approaches (Quenault Béatrice, 2014, p. 92).

In conclusion, although much progress has been made in developing frameworks and tools for assessing urban resilience, significant challenges remain. A promising approach could be to combine quantitative and qualitative methods, to develop indicators that are adaptable to local contexts while allowing broader comparisons, and to adopt a dynamic, multi-scale perspective in assessing urban resilience. These efforts are essential to improve our understanding of urban resilience and to effectively guide policies and interventions aimed at strengthening the capacity of cities to cope with current and future challenges.

## Conclusion:

The theoretical conceptualization and operationalization of urban resilience as a multidimensional paradigm reveal a complex and constantly evolving field of study. This chapter has highlighted the many facets of this concept, from its interdisciplinary origins to its contemporary applications in the urban context, as well as the challenges of measuring and assessing it. The emergence of resilience as a key concept in urban studies testifies to a significant paradigmatic shift in the way urban challenges are approached. As pointed out by Quenault Béatrice, Pigeon Patrick, Bertrand Frédéric and Blond Nadège (2011, p. 20), this shift from a vulnerability-centric approach to a resilience perspective reflects a desire to develop more proactive and holistic strategies in the face of growing uncertainties in the urban world. This conceptual evolution is taking place against a backdrop of accelerating globalization and global environmental change, which require cities to have a greater capacity for adaptation and transformation.

An analysis of the various definitions and interpretations of urban resilience has highlighted the richness but also the complexity of this concept. The variability of approaches across disciplines and application contexts underlines the need for a nuanced and contextualized understanding of urban resilience. As argued by Toubin Marie, Lhomme Serge, Diab Youssef, Serre Damien and Laganier Richard (2012, p. 630), the key components of urban resilience - capacity to absorb shocks, adaptability in the face of gradual change, and transformability - form a rich conceptual framework for analyzing and strengthening the resilience of urban systems.

The conceptualization of urban resilience as a dynamic, multiscalar process has highlighted the importance of interactions between different temporal and spatial scales. This perspective, developed in particular by Lhomme Serge, Serre Damien, Diab Youssef and Laganier Richard (2013, p. 14), highlights the need for a systems approach that takes into account the complex interdependencies within and between urban systems, from the local to the global level.

Exploring the economic, social, environmental and institutional dimensions of urban resilience has revealed the multidimensional nature of this concept. These dimensions, far from being isolated, interact in complex ways, creating synergies but also sometimes tensions. This complexity, highlighted by Rufat Samuel (2012, p. 205), underlines the importance of an integrated and balanced approach to planning and managing urban resilience.

Finally, the challenges of measuring and

assessing multidimensional urban resilience, discussed in particular by Rebotier Julien (2012, p. 393), highlight the current limits of our ability to quantify and compare resilience between different urban contexts. These challenges call for continued development of assessment methodologies and tools, as well as critical reflection on the relevance and limits of quantitative approaches in the study of complex urban phenomena.

In conclusion, this chapter has demonstrated that urban resilience, as a multidimensional paradigm, offers a rich conceptual framework for understanding and strengthening the capacity of cities to face the challenges of the 21st century. However, its effective operationalization requires an interdisciplinary approach, consideration of contextual specificities, and integration of the different dimensions and scales of urban resilience. Future research in this field will need to continue refining our theoretical understanding while developing practical tools to translate this concept into concrete actions aimed at creating more resilient and sustainable cities.

## *Chapter 2: Economic globalization and challenges for Moroccan cities*

Economic globalization, a multidimensional and complex phenomenon, has profoundly reconfigured urban dynamics on a planetary scale. Far from being spared these changes, Moroccan cities are at the heart of major transformations that are redefining their role and place in the national and international economy. The aim of this chapter is to analyze the many facets of the impact of globalization on Moroccan urban areas, highlighting both the opportunities generated and the considerable challenges facing these cities.

Morocco's growing integration into the global economy has led to a spatial and functional reconfiguration of the country's cities. The influx of foreign direct investment, the creation of free trade zones and industrial parks, and the expansion of the tertiary sector have all contributed to the emergence of urban hubs aspiring to international competitiveness. However, this dynamic has been accompanied by far-reaching economic restructuring, marked by partial deindustrialization and job insecurity in certain sectors, raising the question of the economic resilience of urban fabrics.

At the same time, Moroccan cities are facing intense demographic pressures, fueled by sustained natural growth, persistent rural exodus and international migration flows. These demographic dynamics exacerbate pre-existing socio-spatial inequalities, manifesting themselves in increased residential segregation and difficulties of access to housing and urban services for a significant proportion of the urban population.

Finally, Morocco's rapid and sometimes poorly controlled urbanization raises critical environmental issues. Air pollution, waste management, water stress and vulnerability to natural hazards and climate change are all major challenges for the sustainability and resilience of Moroccan cities.

This chapter takes an in-depth look at these issues, drawing on empirical data and theoretical frameworks from urban economics, geography and sociology. The aim is to provide a nuanced picture of the urban transformations brought about by globalization in Morocco, highlighting the interconnections between the economic, social and environmental dynamics at work in the country's cities.

## 2.1 Dynamics of globalization and their impact on urban spaces

### 2.1.1 Morocco's integration into the global economy

Morocco's integration into the global economy has accelerated considerably since the 1980s, marking a major turning point in the country's economic development strategy. This openness has manifested itself through a series of structural reforms and trade agreements that have gradually anchored the Kingdom in international economic flows. According to

Mohammed Abdelmoumni (2019, p. 78), "the process of integrating Morocco into the global economy has been structured around three main axes: liberalizing trade, modernizing the regulatory framework, and attracting foreign direct investment". This multidimensional approach has enabled Morocco to diversify its trading partners and increase its presence on international markets.

Morocco's accession to the World Trade Organization (WTO) in 1995 marked a crucial step in this integration process. This event not only opened up new commercial opportunities for the country, but also required national economic policies to be adapted to international standards. As Nadia El Hachimi (2021, p. 132) points out, "accession to the WTO catalyzed a series of reforms aimed at improving the competitiveness of the Moroccan economy, particularly in the industrial and services sectors". These reforms have included modernizing the tax system, simplifying administrative procedures for businesses, and strengthening intellectual property protection.

At the same time, Morocco has actively pursued a policy of signing bilateral and multilateral free-trade agreements. The Association Agreement with the European Union, which came into force in 2000, has been particularly significant, gradually opening up the Moroccan market to European products while facilitating access for Moroccan exports to the European market. Youssef El Aloui (2020, p. 215) argues that "these agreements have not only stimulated trade, but have also encouraged the transfer of technology and know-how, thus contributing to the modernization of Morocco's productive apparatus".

However, this increased integration into the global economy is not without its challenges. Karima Ghazouani (2022, p. 56) highlights "the increased vulnerability of the Moroccan economy to external shocks, notably fluctuations in raw material prices and international financial crises". This exposure has necessitated the development of economic resilience mechanisms and the diversification of export sectors to reduce dependence on a limited number of markets or products.

Economic integration has also had a profound impact on Morocco's urban fabric. Cities, particularly metropolises such as Casablanca, Tangier and Marrakech, have become nodal points for this insertion into the global economy. Hassan Zaoual (2018, p. 189) observes that "economic openness has accelerated the transformation of major Moroccan cities into poles of attraction for foreign investment, thereby modifying their economic and social structure". This dynamic has given rise to new challenges in terms of urban planning, managing internal migratory flows, and balancing economic development with the preservation of cultural heritage.

In conclusion, Morocco's integration into the global economy is a multifaceted process that has profoundly reconfigured the country's economic and urban landscape. While bringing significant opportunities in terms of growth and modernization, this integration also raises crucial questions about the country's ability to maintain a balanced and inclusive development trajectory in the face of the pressures of international competition.

## 2.1.2 Foreign direct investment flows to Moroccan cities

Foreign direct investment (FDI) flows to Moroccan cities have grown significantly in recent decades, reflecting Morocco's growing attractiveness as a destination for international capital. This phenomenon has had a profound influence on the country's economic and urban development, with varying impacts depending on the region and sector of activity.

According to Rachid El Houdaigui (2021, p. 87), "the dynamics of FDI in Morocco have been characterized by a marked geographical concentration, with a predominance of investment in major urban centers, notably Casablanca, Tangier and Rabat". This trend can be explained by the presence in these cities of well-developed infrastructures, a pool of skilled labor, and a business environment more conducive to international activities. Casablanca, in particular, has established itself as the country's main financial and economic hub, attracting a substantial share of FDI in the services, finance and manufacturing sectors.

Tangier's emergence as a major pole of attraction

for FDI illustrates the transformative impact these flows can have on urban development. Nadia Benabdeljlil (2020, p. 143) points out that "the establishment of the Tangier Med port complex and the creation of adjacent free zones have radically altered the city's economic profile, attracting massive investment in the automotive, aeronautics and logistics sectors". This dynamic has not only stimulated job creation, but has also led to a spatial reconfiguration of the region, with the emergence of new residential and commercial zones.

However, the distribution of FDI across Morocco remains uneven. As Mohammed Tozy (2019, p. 210) notes, "despite economic decentralization efforts, inland cities and outlying regions are still struggling to attract a significant share of foreign investment". This disparity raises questions about the balance of territorial development and the risks of accentuating regional inequalities.

The sectors benefiting from FDI in Moroccan cities have evolved over time, reflecting changes in the global economy and Morocco's strategic priorities. Fatima Zohra Saïdi (2022, p. 168) observes that "while manufacturing and tourism have long dominated FDI flows, the last decade has seen a diversification towards higher value-added sectors such as renewable energies, information technology and offshore services". This evolution has important implications for skills development and the structuring of the urban labor market.

The impact of FDI on Morocco's urban fabric is multi-faceted. On the one hand, they have contributed to infrastructure modernization, job creation, and the transfer of technologies and managerial practices. On the other hand, as Youssef Courbage (2018, p. 95) points out, "the influx of FDI has accentuated certain gentrification and spatial segregation dynamics in major cities, with the emergence of high-end business districts and residential areas that are sometimes disconnected from the traditional urban fabric".

The Moroccan authorities have put in place various measures to optimize the impact of FDI on sustainable urban development. Hassan Zaoual (2020, p. 276) mentions "the elaboration of urban development master plans integrating foreign investment projects into a coherent vision of territorial development". These efforts aim to reconcile economic attractiveness, urban quality of life and preservation of the cultural identity of cities.

In conclusion, FDI flows to Moroccan cities have played a catalytic role in their economic and spatial transformation. While offering significant development opportunities, they also pose challenges in terms of territorial equity and urban social cohesion. The strategic management of these flows and their harmonious integration into the urban fabric remain crucial issues for the sustainable development of Moroccan cities.

### 2.1.3 Development of free zones and industrial parks

The development of free zones and industrial parks in Morocco represents a key strategy in the country's efforts to boost industrialization and integration into the global economy. These specialized economic areas have profoundly altered the country's industrial and urban landscape, creating new growth poles and redefining territorial dynamics.

According to Abdelkader Kaioua (2019, p. 123), "the policy of creating free zones in Morocco is part of a rationale of territorial competitiveness aimed at attracting foreign investment and stimulating exports". This approach has led to the emergence of several major free zones, the most emblematic of which is Tangier Med. Kaioua points out that "these economic enclaves benefit from preferential tax and customs regimes, as well as state-of-the-art infrastructures, creating an environment that is conducive to the establishment of international companies".

The impact of these free zones on urban development is significant. Fatima-Zahra Alaoui (2021, p. 87) observes that "the establishment of the Tangier Med free zone catalyzed a rapid urbanization of the region, with the emergence of new residential, commercial and service areas to meet the needs of companies and their employees". This phenomenon illustrates how free zones can act as engines of urban transformation, generating ripple effects throughout the local economic and social fabric.

Alongside free zones, Morocco has also focused on the development of integrated industrial parks. Hassan El Mokri (2020, p. 201) explains that "these parks aim to create coherent industrial ecosystems, bringing together complementary companies within the same value chain". This approach has been particularly visible in sectors such as automotive and aeronautics, where specialized parks have emerged around major international prime contractors.

However, the development of these specialized economic areas also raises challenges. Nadia Benabdellil (2022, p. 156) warns of "the risk of creating economic enclaves disconnected from the local productive fabric, limiting the spin-offs in terms of technology transfer and skills development for the national economy". This concern underlines the importance of further integrating free zones and industrial parks into overall territorial development strategies.

The environmental dimension is also crucial in assessing the impact of these zones. Youssef El Jai (2021, p. 289) notes that "the concentration of industrial activities in these areas poses challenges in terms of resource management, waste treatment and pollution". He stresses the growing importance of sustainability criteria in the design and management of new industrial zones in Morocco.

In terms of urban planning, the integration of

these zones into the existing urban fabric is a major challenge. Mohammed Idrissi Janati (2020, p. 178) observes that "the creation of free zones and industrial parks on the outskirts of cities contributes to urban sprawl and can accentuate socio-spatial disparities". This dynamic raises questions about the coherence of urban planning policies and the need for integrated planning that takes into account the economic, social and environmental aspects of urban development.

In conclusion, the development of free zones and industrial parks in Morocco has played a crucial role in the country's strategy to industrialize and attract foreign investment. While creating new economic opportunities and stimulating the development of certain regions, these areas also pose challenges in terms of urban integration, territorial equity and environmental sustainability. Managing these issues will be decisive in ensuring that these growth poles contribute positively and sustainably to the development of Moroccan cities.

## 2.1.4 Growth of the tertiary sector and business services

The rise of the tertiary sector and business services in Morocco is a major component of the country's economic transformation, reflecting a global trend towards the tertiarization of developing economies. This phenomenon has profound implications for the structure of employment, the spatial organization of cities and the dynamics of urban development.

According to Rachid Boutti (2021, p. 145), "the rise of the tertiary sector in Morocco is part of a process of economic diversification and modernization of the country's productive fabric". This development is characterized by significant growth in service activities, both in the field of personal and business services. Boutti points out that "this tertiarization process has accelerated since the 2000s, driven by economic liberalization policies and the country's increased openness to foreign investment".

The emergence of business services as a key sector of the Moroccan economy is particularly noteworthy. Nadia El Fassi (2020, p. 213) observes that "the development of business services, such as consulting, engineering, financial and legal services, has been stimulated by the increasing complexity of the business environment and the growing internationalization of the Moroccan economy". This dynamic has contributed to the creation of skilled jobs in major cities, making them more attractive to young graduates and professionals.

The spatial impact of this tertiarization is particularly visible in major metropolises. Hassan Radoine (2019, p. 178) notes that "the boom in services has led to a reconfiguration of urban space, with the emergence of modern business districts and the conversion of industrial zones into tertiary hubs". This phenomenon is particularly marked in Casablanca, with the development of zones such as Casablanca Finance City, designed to position the city as a regional financial hub.

However, the concentration of tertiary activities in major urban centers raises questions of territorial balance. Mohammed Tozy (2022, p. 267) warns of "the risk of accentuating regional disparities, with medium-sized and small towns struggling to develop a dynamic tertiary sector capable of retaining local talent". This concern underlines the importance of balanced territorial development policies, aimed at spreading the benefits of tertiary sector growth beyond the main metropolises.

The boom in business services also has important implications in terms of training and skills development. Fatima Zohra Saïdi (2020, p. 132) points out that "the growing demand for qualified profiles in business services has stimulated an adaptation of university and vocational curricula, with an increased emphasis on cross-disciplinary and technological skills". This trend is helping to improve the match between training provision and the needs of the urban job market.

At the same time, the development of the service sector brings with it new challenges in terms of work organization and job quality. Youssef El Aloui (2021, p. 189) notes that "while the service sector offers opportunities for skilled employment, it is also characterized by a growing casualization of certain segments, particularly in support activities and low value-added services". This duality of the tertiary labor market raises questions about social cohesion in urban areas and the ability of cities to generate inclusive development.

Technological innovation is playing a crucial role in the transformation of the Moroccan service sector. Karim Benabdallah (2022, p. 245) observes that "the digitization of services and the emergence of the digital economy are redefining business models and the skills required, offering new opportunities but also posing challenges in terms of adapting the workforce and urban infrastructures".

In conclusion, the rise of the tertiary sector and business services in Morocco represents a profound transformation of the urban economy, with significant implications for spatial organization, the labor market and city development dynamics. While offering opportunities for growth and modernization, this evolution raises important issues in terms of territorial equity, training and social inclusion. The ability of Moroccan cities to manage these challenges and capitalize on the opportunities offered by tertiarization will be decisive for their future development and international competitiveness.

## 2.1.5 Emergence of internationally competitive urban clusters

The emergence of internationally competitive urban hubs in Morocco bears witness to a profound transformation of the country's urban and economic landscape. This phenomenon, resulting from a convergence of economic, political and strategic factors, has considerably reconfigured the national urban hierarchy and positioned certain Moroccan cities on the world stage of attractive metropolises.

According to Mohammed Berriane (2019, p. 112), "the emergence of competitive urban clusters in Morocco is part of a national strategy to create 'locomotive cities' capable of stimulating regional development and inserting themselves into global economic networks". This approach has led to the concentration of massive investment in certain conurbations, transforming their physiognomy and economic structure.

Casablanca remains the country's main economic hub, but its international positioning has been considerably strengthened. As Nadia El Hachimi (2021, p. 178) points out, "the Casablanca Finance City project illustrates the city's determination to become a regional financial hub, capable of competing with the likes of Dubai and Singapore". This ambitious project is accompanied by a modernization of urban infrastructures and an improvement in the business environment, aimed at attracting the regional headquarters of multinationals and international financial institutions.

Tangier is an emblematic case of the rapid emergence of a competitive urban cluster. Hassan

Zaoual (2020, p. 245) observes that "Tangier's transformation into a world-class logistics and industrial hub, centered around the Tangier Med port complex, has radically altered the city's position in the global economy". This transformation has been accompanied by a profound reconfiguration of urban and suburban space, with the emergence of new economic and residential centralities.

Rabat, the administrative capital, is also undergoing a transformation aimed at boosting its international appeal. Fatima Zohra Saïdi (2022, p. 201) notes that "the 'Rabat Ville Lumière, Capitale Marocaine de la Culture' project aims to position the city as a major cultural and diplomatic hub, complementing Casablanca's economic functions". This strategy illustrates the diversification of approaches to creating competitive urban clusters, capitalizing on specific assets beyond the purely economic dimension.

The emergence of these competitive urban hubs brings with it major challenges. Youssef El Aloui (2021, p. 167) warns of "the risk of accentuating territorial disparities, with the concentration of investments and opportunities in a limited number of metropolises to the detriment of medium-sized and small towns". This concern raises the question of balanced territorial development and the need for effective redistribution mechanisms.

Environmental sustainability is another crucial

issue. Karim Benabdallah (2020, p. 289) points out that "the race for international competitiveness must not be at the expense of urban quality of life and the preservation of natural resources". This perspective highlights the importance of integrating sustainability criteria into competitive urban development strategies.

The insertion of these urban hubs into international networks also raises questions of cultural identity and social cohesion. Mohammed Idrissi Janati (2019, p. 223) observes that "the rapid internationalization of certain neighborhoods sometimes creates tensions with the traditional urban fabric and can give rise to phenomena of gentrification and exclusion". Managing these socio-spatial dynamics represents a major challenge for urban authorities.

In conclusion, the emergence of internationally competitive urban clusters in Morocco reflects a national ambition to position itself in the globalized economy. This process, which offers opportunities in terms of economic growth and attractiveness, also raises crucial issues in terms of territorial equity, environmental sustainability and social cohesion. The ability to reconcile international competitiveness and inclusive urban development will be a determining factor in the future trajectory of these emerging Moroccan metropolises.

## 2.2 Economic restructuring, deindustrialization and urban insecurity

## 2.2.1 Changes in the traditional industrial fabric

Changes in Morocco's traditional industrial fabric represent a complex, multidimensional phenomenon, reflecting the profound transformations of the national economy in the context of globalization. This process, which has accelerated since the 1990s, has profoundly reconfigured the country's industrial landscape, with significant implications for cities and regions.

According to Hassan El Mokri (2019, p. 156), "the restructuring of Morocco's industrial fabric has been characterized by a gradual transition from traditional, often low value-added industries to more technological and internationally competitive sectors". This evolution has been particularly marked in sectors such as textiles, agro-industry and mechanical engineering, which have had to reinvent themselves in the face of growing international competition.

Fatima Zohra Saïdi (2021, p. 213) points out that "the modernization of the traditional industrial fabric has been stimulated by a combination of factors, including trade openness, industrial upgrading policies, and the influx of foreign direct investment". These dynamics have led to a spatial reconfiguration of industry, with the decline of certain historical industrial basins and the emergence of new specialized poles.

The impact of these changes on the urban fabric is significant. Mohammed Tozy (2020, p. 178) observes that "the partial de-industrialization of certain urban areas, notably in the former industrial peripheries of

Casablanca or Fez, has created economic and social reconversion challenges". This phenomenon raises crucial questions about the ability of cities to manage these transitions and maintain their economic dynamism.

At the same time, new industrial sectors with higher added value are emerging. Nadia Benabdellil (2022, p. 245) notes that "the boom in the automotive and aeronautics industries, particularly around Tangiers and Casablanca, illustrates Morocco's ability to insert itself into more sophisticated global value chains". This development is accompanied by new needs in terms of skills, infrastructure and the spatial organization of production.

However, these changes are not without social consequences. Youssef El Aloui (2021, p. 189) warns of "the risk of deskilling and casualization of part of the traditional industrial workforce, ill-prepared for the demands of the new emerging sectors". This concern underlines the importance of support and training policies to facilitate the transition of the workforce.

The environmental dimension is also crucial to the analysis of these changes. Karim Benabdallah (2020, p. 301) points out that "industrial modernization offers opportunities for cleaner, more efficient production, but also raises challenges in terms of brownfield management and the reconversion of polluted sites". This perspective highlights the need to integrate environmental considerations into industrial restructuring strategies.

The impact of digitalization and Industry 4.0 on the traditional industrial fabric is an emerging issue. Rachid Boutti (2022, p. 167) observes that "the adoption of digital technologies and automation is redefining production methods and the skills required, posing adaptation challenges for traditional industries". This technological transition brings with it new opportunities, but also risks creating a digital divide between companies and regions.

In conclusion, changes in Morocco's traditional industrial fabric reflect a complex process of modernization and adaptation to the challenges of global competitiveness. These transformations, while offering opportunities to move upmarket and integrate into the global economy, raise crucial issues in terms of territorial cohesion, environmental sustainability and social inclusion. The ability of cities and regions to manage these transitions, capitalizing on their specific assets while mitigating negative impacts, will be decisive for Morocco's industrial future.

### 2.2.2 Plant closures and job losses in certain sectors

Plant closures and job losses in certain industrial sectors in Morocco are a complex and worrying phenomenon, reflecting the challenges facing the country in a context of globalization and economic restructuring. This dynamic has profound implications for the social and economic fabric of Moroccan cities, particularly those with a strong industrial tradition.

According to Mohammed Abdelmoumni (2020, p. 178), "the wave of factory closures observed since the early 2000s has mainly affected traditional sectors of Moroccan industry, such as textiles, footwear and certain branches of the agro-industry". This phenomenon can be largely explained by the intensification of international competition, particularly from Asian countries, and by the inability of certain companies to adapt to the new demands of the global market.

Fatima El Hassouni (2021, p. 213) points out that "job losses linked to factory closures have particularly affected the cities of Casablanca, Fez and Tangier, historically bastions of the textile and manufacturing industry". This situation has led to pockets of structural unemployment in certain neighborhoods, exacerbating social tensions and the challenges of economic inclusion in these urban areas.

The impact of these closures on the urban fabric is significant. Nadia Benabdellil (2019, p. 145) observes that "the partial deindustrialization of certain urban areas has led to the emergence of industrial wastelands, posing major challenges in terms of urban redevelopment and economic conversion". This phenomenon raises crucial questions about the ability of cities to reinvent themselves and maintain their economic dynamism in the face of these changes.

However, it is important to note that these closures are part of a broader context of industrial restructuring. Youssef El Aloui (2022, p. 267) argues that "in parallel with the closures in traditional sectors, we are seeing the emergence of new, higher value-added industries, notably in the automotive and aeronautics sectors". This dynamic illustrates the complexity of the changes underway, combining the decline of certain sectors with the emergence of new growth poles.

The social consequences of these closures are particularly worrying. Hassan Zaoual (2020, p. 189) highlights "the risk of casualization of part of the industrial workforce, often poorly qualified and ill-prepared for retraining in emerging sectors". This situation underscores the urgent need to implement support and training policies to facilitate the professional transition of affected workers.

The gendered dimension of these job losses deserves particular attention. Karima Ghazouani (2021, p. 233) notes that "women, over-represented in certain sectors such as textiles, are particularly vulnerable to factory closures, which can have significant repercussions on household economic balance and gender dynamics within urban communities".

Faced with these challenges, the Moroccan authorities have introduced a number of initiatives. Rachid Boutti (2022, p. 301) points out that "industrial conversion support and vocational training programs have been launched to mitigate the impact of plant closures and facilitate the reintegration of workers".

However, the effectiveness of these measures remains debated, particularly in terms of their ability to reach the most vulnerable populations.

The environmental implications of factory closures should not be overlooked. Mohammed Idrissi Janati (2020, p. 178) observes that "the closure of industrial sites poses challenges in terms of land decontamination and rehabilitation, requiring significant investment and integrated urban planning".

In conclusion, plant closures and job losses in certain industrial sectors in Morocco represent a major challenge for urban development and social cohesion. Although part of a broader process of economic restructuring, these phenomena raise crucial questions in terms of social justice, urban resilience and environmental sustainability. The ability of Moroccan cities to manage these transitions, minimizing negative impacts while seizing opportunities for reconversion and innovation, will be decisive for their future development trajectory.

## 2.2.3 Precarious employment and development of the informal sector

Job insecurity and the development of the informal sector in Morocco are intrinsically linked phenomena that have grown significantly in recent decades. These dynamics, the result of multiple economic and social factors, have had a profound impact on the urban fabric and living conditions in Moroccan cities.

According to Noureddine El Aoufi (2020, p. 156), "the casualization of employment in Morocco is reflected in an increase in fixed-term contracts, temporary work and involuntary part-time work". This trend, particularly marked in urban areas, reflects changes in the labor market in the face of the pressures of international competitiveness and economic flexibilization.

At the same time, the informal sector has expanded considerably. Fatima Zohra Saïdi (2021, p. 213) observes that "the informal economy now accounts for almost 30% of Moroccan GDP and employs a significant proportion of the urban workforce". This phenomenon, while offering a safety valve in the face of unemployment, raises crucial questions in terms of social protection, working conditions and taxation.

The spatial impact of these dynamics is particularly visible in large cities. Mohammed Tozy (2019, p. 178) notes that "the casualization of employment and informality have led to a reconfiguration of urban space, with the emergence of zones of informal activity and the multiplication of street markets". This development poses major challenges in terms of urban planning and public space management.

The gendered dimension of these phenomena deserves particular attention. Rajaa Mejjati Alami (2022, p. 245) points out that "women are over-represented in precarious employment and the informal sector, particularly in activities such as domestic work and small-scale street trading". This situation exacerbates gender inequalities and compromises the economic empowerment of urban women.

The link between job insecurity and the development of the informal sector is complex. Hassan El Houari (2021, p. 189) argues that "informality often appears as a survival strategy in the face of the precariousness of the formal labor market, but it can also constitute a trap preventing access to more stable, better-protected jobs". This dynamic creates a vicious circle that is difficult to break for many urban workers.

The social consequences of these phenomena are manifold. Youssef El Aloui (2020, p. 301) highlights "the erosion of social protection, the increase in household economic vulnerability and the weakening of social cohesion in urban neighborhoods". These trends pose major challenges in terms of social policy and sustainable urban development.

Faced with these challenges, the Moroccan authorities have launched a number of initiatives. Karim Benabdallah (2021, p. 167) notes that "programs aimed at gradually formalizing the informal economy and extending social coverage have been launched, but their effectiveness remains limited given the scale of the phenomenon". These efforts underline the complexity of striking a balance between economic flexibility and social protection.

The impact of digitalization on these dynamics also deserves attention. Nadia Benabdellil (2022, p. 223) observes that "the emergence of the platform economy and on-demand work introduces new forms of precariousness, while offering flexible employment opportunities in urban areas". This development raises new questions about labor regulation and worker protection in the digital economy.

In conclusion, precarious employment and the development of the informal sector in Morocco represent major challenges for urban development and social cohesion. These phenomena, while offering a degree of economic flexibility and opportunities for survival, raise crucial questions in terms of equity, social protection and the sustainability of the urban development model. The ability of Moroccan cities to manage these dynamics, striking a balance between flexibility and security, will be decisive for their future trajectory and the well-being of their inhabitants.

## 2.2.4 Growing socio-economic inequalities within cities

The increase in intra-urban socio-economic inequalities in Morocco is a complex, multidimensional phenomenon that has become more pronounced in recent decades. This dynamic reflects the profound transformations of the Moroccan economy and society in a context of globalization and urban restructuring.

According to Mohammed Idrissi Janati (2021, p. 178), "the sustained economic growth experienced by

Morocco since the early 2000s has been accompanied by increased social polarization within major cities". This observation highlights the paradox of an economic development that, while generating wealth, has not necessarily led to a reduction in social disparities in urban areas.

Fatima El Hassouni (2020, p. 213) points out that "the increase in intra-urban inequalities is reflected in growing spatial segregation, with the emergence of affluent neighborhoods alongside areas of precarious or informal housing". This fragmentation of urban space reflects and reinforces socio-economic disparities, creating "cities within cities" with often contrasting realities.

Unequal access to basic urban services is a crucial dimension of these inequalities. Nadia Benabdellil (2022, p. 145) observes that "disparities in terms of access to education, health, transport and green spaces are particularly marked between the different districts of Morocco's major metropolises". These inequalities in access to urban resources compromise equality of opportunity and perpetuate cycles of intergenerational poverty.

The gendered dimension of these inequalities deserves particular attention. Rajaa Mejjati Alami (2019, p. 267) highlights that "women, particularly in disadvantaged neighborhoods, are often harder hit by socio-economic inequalities, facing additional obstacles in terms of access to formal employment and urban services". This situation underlines the importance of integrating a gender perspective into urban development policies.

The urban labor market plays a central role in the reproduction and amplification of inequalities. Hassan El Houari (2021, p. 189) argues that "the polarization of the labor market between highly-skilled, well-paid jobs on the one hand, and precarious or informal jobs on the other, contributes to widening the gap between different urban socio-professional categories". This dynamic poses major challenges in terms of social cohesion and economic mobility.

The impact of urban renewal policies and major projects on inequalities is ambivalent. Youssef El Aloui (2020, p. 301) notes that "while these initiatives often aim to improve the attractiveness and competitiveness of cities, they can also lead to gentrification and the displacement of the most vulnerable populations". This tension between urban development and social equity raises crucial questions about urban governance and planning.

In response to these challenges, the Moroccan authorities have launched a number of initiatives. Karim Benabdallah (2022, p. 156) points out that "programs have been launched to reduce substandard housing, improve access to basic services and promote employment in disadvantaged neighborhoods". However, the effectiveness of these measures remains debated, particularly in terms of their ability to tackle the structural causes of inequality.

The issue of urban mobility is also central to the problem of inequality. Noureddine El Aoufi (2020, p. 223) observes that "disparities in terms of access to public transport and travel time between the center and the periphery reinforce inequalities in access to employment and economic opportunities". This dimension underlines the importance of integrated transport and urban planning to promote a more inclusive city.

In conclusion, the increase in intra-urban socio-economic inequalities in Morocco represents a major challenge for the sustainable development and social cohesion of cities. This phenomenon, the result of complex dynamics linked to economic growth, urban restructuring and changes in the labor market, calls for integrated, multidimensional approaches. The ability of Moroccan cities to reduce these disparities, by promoting more inclusive and equitable development, will be decisive for their future and the well-being of all their inhabitants.

## 2.2.5 The challenges of brownfield redevelopment

Brownfield redevelopment in Morocco represents a major challenge for sustainable urban development and the economic revitalization of cities. This phenomenon, resulting from changes in the industrial fabric and the dynamics of partial deindustrialization, raises complex issues in terms of urban planning, environmental sustainability and local economic development.

According to Hassan Radoine (2020, p. 178), "industrial wastelands, particularly present in the former industrial zones of Casablanca, Mohammedia and Fez, are both a problematic legacy and an opportunity for urban reinvention". This duality underlines the need for a strategic and innovative approach to the management of these areas.

The environmental dimension of brownfield redevelopment is crucial. Mohammed Berriane (2021, p. 213) observes that "soil and groundwater pollution in these areas poses major challenges in terms of decontamination and ecological rehabilitation". This situation calls for substantial investment and specific technical expertise, which are often difficult for local authorities to mobilize.

The economic implications of conversion are also central. Fatima Zohra Saïdi (2019, p. 145) points out that "the transformation of brownfield sites into new hubs of economic activity can play a key role in the revitalization of declining urban districts". However, such reconversion requires a clear vision of local economic development and the ability to attract new investors and activities.

The heritage dimension should not be overlooked in this process. Nadia El Fassi (2022, p. 267) argues that "certain industrial wastelands, which bear witness to Morocco's industrial history, deserve to be preserved and developed from the perspective of industrial tourism and working-class memory". This approach makes it possible to combine economic reconversion with the preservation of urban cultural heritage.

The governance challenges associated with brownfield redevelopment are considerable. Youssef El Aloui (2021, p. 189) highlights "the complexity of the legal and financial arrangements required for redevelopment, often involving a multiplicity of public and private players". This situation underlines the importance of effective coordination between the various stakeholders and of an appropriate regulatory framework.

The social impact of brownfield redevelopment deserves particular attention. Rajaa Mejjati Alami (2020, p. 301) notes that "the transformation of these spaces can give rise to gentrification phenomena, raising the question of the inclusion of local populations in reconversion projects". This concern underlines the need to integrate a strong social dimension into reconversion strategies.

In response to these challenges, a number of innovative approaches have emerged. Karim Benabdallah (2022, p. 156) observes that "some Moroccan cities are experimenting with reconversion models based on the creative economy and innovation, transforming former factories into coworking spaces, incubators or cultural venues". These initiatives illustrate the potential for urban reinvention offered by brownfield sites.

The question of functional mix is central to redevelopment projects. Noureddine El Aoufi (2021, p. 223) points out that "the integration of residential, commercial and recreational uses in redevelopment projects makes it possible to create dynamic, multifunctional new neighborhoods". This approach contributes to the creation of livelier, more sustainable urban spaces.

The issue of mobility and accessibility is also crucial in the redevelopment of brownfield sites. Hassan El Houari (2020, p. 178) notes that "the reintegration of these often peripheral spaces into the urban fabric requires in-depth consideration of transport connections and links with the rest of the city". This dimension underlines the importance of an integrated approach to urban planning.

In conclusion, brownfield redevelopment in Morocco represents a complex challenge, but also a major opportunity for urban renewal. This process requires a multidimensional approach, integrating environmental, economic, social and heritage considerations. The ability of Moroccan cities to transform these areas into new hubs of urban dynamism will be decisive for their future development and their ability to meet the challenges of sustainability and social inclusion. The success of these reconversion projects could serve as a model for other cities facing similar challenges in the region.

## 2.3 Demographic pressures, migration and socio-

**spatial inequalities**

## 2.3.1 Population growth and urban sprawl

Population growth and urban sprawl in Morocco are closely linked phenomena that have had a profound impact on the development of Moroccan cities over the last few decades. These dynamics have major implications for spatial organization, the environment and urban resource management.

According to Mohammed Idrissi Janati (2021, p. 178), "the sustained demographic growth of Moroccan cities, fuelled by natural increase and rural exodus, has led to an unprecedented spatial expansion of urban areas". This observation highlights demographic pressure as the main driver of urban sprawl. The Haut-Commissariat au Plan (2020) estimates that Morocco's urban population will more than double between 1980 and 2020, rising from 8.7 million to over 22 million.

Fatima El Hassouni (2020, p. 213) points out that "urban sprawl in Morocco is characterized by the horizontal extension of cities, often to the detriment of peri-urban agricultural land". This dynamic poses major challenges in terms of environmental sustainability and food security, particularly around major metropolises such as Casablanca, Rabat and Marrakech.

The impact of this sprawl on urban infrastructures is considerable. Nadia Benabdellil (2022, p. 145) observes that "the rapid expansion of urban peripheries is putting pressure on transport, sanitation and water and electricity supply networks". This situation raises crucial questions about the ability of cities to provide quality basic services to an ever-expanding urban population.

The social dimension of urban sprawl deserves particular attention. Hassan El Houari (2021, p. 189) points out that "urban expansion is often accompanied by increased socio-spatial segregation, with the emergence of poorly integrated and under-equipped peripheral neighborhoods". This dynamic poses major challenges in terms of social cohesion and equity of access to urban opportunities.

The role of land speculation in urban sprawl is underlined by Youssef El Aloui (2020, p. 301), who argues that "rising land prices in city centers push low-income populations to increasingly distant outskirts, thus accelerating sprawl". This situation calls for more effective regulation of the urban land market and more inclusive social housing policies.

In response to these challenges, the Moroccan authorities have launched a number of initiatives. Karim Benabdallah (2022, p. 156) notes that "programs for new towns and integrated urban poles have been launched to channel urban growth and offer alternatives to anarchic sprawl". However, the effectiveness of these measures remains debated, particularly in terms of their ability to create lively, attractive urban spaces.

The issue of mobility is central to the problem of urban sprawl. Rajaa Mejjati Alami (2019, p. 267) points out that "the extension of urban peripheries increases dependence on the automobile and poses major challenges in terms of congestion and pollution". This situation calls for integrated urban and transport planning, favoring more compact, multimodal urban development models.

The environmental impact of urban sprawl is considerable. Mohammed Berriane (2021, p. 223) observes that "the increasing artificialization of peri-urban soils threatens local biodiversity and increases the vulnerability of cities to natural hazards such as flooding". This concern underlines the importance of integrating strong ecological considerations into urban planning policies.

In conclusion, population growth and urban sprawl in Morocco represent major challenges for the sustainable development of cities. These phenomena, resulting from complex dynamics linked to demographics, economics and urban governance, require integrated, multidimensional approaches. The ability of Moroccan cities to manage this growth sustainably, by promoting more compact, equitable and environmentally-friendly urban development models, will be decisive for their future and the quality of life of their inhabitants.

## 2.3.2 Rural exodus and accelerated urbanization

Morocco's rural exodus and accelerated urbanization are closely interlinked phenomena that have profoundly transformed the country's socio-economic and spatial landscape in recent decades. These dynamics have major implications for urban development, social structure and the national economy.

According to Mohammed Tozy (2020, p. 178), "Morocco's rural exodus has intensified since the 1960s, fueled by growing economic disparities between rural and urban areas, as well as recurrent droughts affecting agriculture". This observation highlights the structural factors driving rural populations towards the cities. The Haut-Commissariat au Plan (2021) estimates that the urbanization rate in Morocco has risen from 29.2% in 1960 to over 63% in 2020, illustrating the scale of this phenomenon.

Fatima Zohra Saïdi (2019, p. 213) points out that "accelerated urbanization has led to explosive population growth in Morocco's major cities, particularly in the Casablanca-Kénitra Atlantic axis". This urban concentration poses major challenges in terms of land use planning and urban infrastructure management.

The impact of the rural exodus on destination cities is considerable. Nadia El Fassi (2022, p. 145) observes that "the massive influx of rural migrants has contributed to the expansion of peripheral and informal settlements, often characterized by a lack of access to basic services and increased precariousness". This situation raises crucial questions about the ability of cities to integrate these new populations and offer them decent living conditions.

The social dimension of the rural exodus deserves particular attention. Hassan El Houari (2021, p. 189) points out that "the uprooting of rural populations and their adaptation to urban life are often accompanied by social and cultural tensions, particularly for the younger generations". This dynamic poses major challenges in terms of social cohesion and urban integration.

The impact of accelerated urbanization on the urban economy is highlighted by Youssef El Aloui (2020, p. 301), who argues that "the influx of rural labor has contributed to the development of the urban informal sector, while exerting downward pressure on wages in certain sectors". This situation calls for appropriate employment and training policies to facilitate the economic integration of rural migrants.

In response to these challenges, the Moroccan authorities have launched a number of initiatives. Karim Benabdallah (2022, p. 156) notes that "programs for rural development and the upgrading of small and medium-sized towns have been launched to mitigate rural exodus and promote more balanced territorial development". However, the effectiveness of these measures remains debated, particularly in terms of their ability to create sustainable economic opportunities in rural areas.

The issue of housing is central to the problem of accelerated urbanization. Rajaa Mejjati Alami (2019, p.

267) points out that "the growing demand for urban housing, fuelled by the rural exodus, has led to a proliferation of informal housing and increased pressure on the formal real estate market". This situation calls for more ambitious social housing policies and more effective regulation of the real estate sector.

The environmental impact of accelerated urbanization is considerable. Mohammed Berriane (2021, p. 223) observes that "the rapid expansion of urban areas to the detriment of agricultural land and natural spaces poses major challenges in terms of environmental sustainability and food security". This concern underlines the importance of urban planning that incorporates strong ecological considerations.

In conclusion, rural exodus and accelerated urbanization in Morocco represent major challenges for the country's sustainable development. These phenomena, resulting from complex dynamics linked to the economy, demographics and public policies, require integrated, multidimensional approaches. Morocco's ability to manage this urban transition in a balanced way, by promoting more inclusive territorial development and improving living conditions in both rural and urban areas, will be decisive for its socio-economic future and national cohesion.

### 2.3.3 International migration and its impact on cities

International migration and its impact on

Moroccan cities is a complex, multi-dimensional phenomenon that has had a considerable influence on the country's urban development over recent decades. This migration dynamic, both inward and outward, has profound implications for the social, economic and spatial fabric of Moroccan cities.

According to Mohamed Berriane (2020, p. 178), "Morocco has gone from being a country of emigration to a country of transit and immigration, particularly for sub-Saharan migrants, transforming the demographic composition of certain cities". This observation highlights the growing complexity of migratory flows and their impact on urban diversity. The Haut-Commissariat au Plan (2021) estimates that around 100,000 sub-Saharan migrants will be living in Morocco in 2020, mainly in large cities.

Fatima El Hassouni (2021, p. 213) points out that "Moroccan emigration to Europe and the Gulf countries has had a significant impact on urbanization, notably through real estate investments by Moroccans living abroad (MREs) in their home towns". This dynamic has contributed to the transformation of the urban landscape, particularly in traditional emigration regions such as the Rif and Souss.

The impact of international migration on the urban economy is considerable. Nadia Benabdellil (2022, p. 145) observes that "remittances from MREs play a crucial role in the economy of certain cities, stimulating consumption and investment in the real estate sector". This has important implications for the housing market and local economic development.

The social dimension of international migration deserves particular attention. Hassan El Houari (2019, p. 189) highlights that "the integration of sub-Saharan migrants in Moroccan cities poses challenges in terms of social cohesion and access to basic services". This dynamic raises crucial questions about the ability of cities to manage cultural diversity and promote social inclusion.

The spatial impact of international migration is highlighted by Youssef El Aloui (2020, p. 301), who argues that "the concentration of certain migrant communities in specific neighborhoods can lead to spatial segregation and gentrification". This situation calls for urban policies that promote social diversity and the integration of migrant populations.

In response to these challenges, the Moroccan authorities have launched a number of initiatives. Karim Benabdallah (2022, p. 156) notes that "migrant regularization and socio-economic integration programs have been launched, aimed at facilitating access to housing, education and employment for migrant populations". However, the effectiveness of these measures remains debated, particularly in terms of their scope and long-term impact.

The question of urban governance in the face of international migration is central. Rajaa Mejjati Alami (2021, p. 267) points out that "Moroccan cities are faced with the need to adapt their policies and services to meet the specific needs of migrant populations, while preserving social cohesion". This situation calls for a more inclusive and participatory approach to urban governance.

The impact of international migration on urban identity is also significant. Mohammed Idrissi Janati (2020, p. 223) observes that "the cultural contribution of international migrants contributes to the diversification and enrichment of the cultural landscape of Moroccan cities, particularly in areas such as gastronomy, music and the arts". This dynamic offers opportunities for the development of cultural tourism and the urban creative economy.

In conclusion, international migration and its impact on Moroccan cities represent both challenges and opportunities for sustainable urban development. These phenomena, resulting from complex dynamics linked to globalization, economic inequalities and individual aspirations, require integrated, multidimensional approaches. The ability of Moroccan cities to manage these migratory flows inclusively and capitalize on the diversity they bring will be decisive for their future and their positioning in an increasingly interconnected world.

## 2.3.4 Residential segregation and fragmentation of urban space

Residential segregation and fragmentation of urban space have become prominent features of

Moroccan cities in the context of globalization and rapid urbanization. These phenomena reflect and reinforce growing socio-economic inequalities within urban areas, posing major challenges for social cohesion and sustainable urban development.

Residential segregation manifests itself in the uneven spatial distribution of social groups within the city, leading to the formation of socio-economically homogeneous neighborhoods. This process is fueled by a variety of factors, including real estate market mechanisms, urban policies, and the residential preferences of different social groups. As Mohammed Aderghal (2015, p. 87) explains in his analysis of the Casablanca metropolis, "socio-spatial segregation has increased with the emergence of new closed residential spaces for the affluent classes, contrasting sharply with persistent working-class neighborhoods and slums". This dynamic contributes to widening gaps between different parts of the city in terms of access to urban services, economic opportunities and quality of life.

At the same time, the fragmentation of urban space is characterized by a physical and functional fragmentation of the city into distinct and often disconnected entities. This phenomenon is particularly visible in Morocco's major conurbations, where urban sprawl and the development of new peripheral economic hubs have led to a polycentric, discontinuous urban structure. Aziz Iraki (2018, p. 142) points out that "urban fragmentation in Morocco results from a combination of factors, including land speculation, major urban projects disconnected from the existing fabric, and the absence of integrated urban planning on a metropolitan scale".

These dynamics of segregation and fragmentation have profound implications for the functioning of Moroccan cities and the well-being of their inhabitants. They exacerbate inequalities in access to economic opportunities and urban services, limit social interaction between different groups, and can fuel social tensions. What's more, they pose considerable challenges in terms of urban governance and the management of public services. As Hind Bouzekri (2020, p. 209) notes, "urban fragmentation makes it more difficult to implement coherent urban policies and to deliver basic services equitably to the entire urban population".

Faced with these challenges, urban policies in Morocco have begun to incorporate approaches aimed at promoting greater social mixing and integration of different parts of the city. However, as Abderrahmane Rachik (2017, p. 176) points out, "efforts to counter urban segregation and fragmentation come up against powerful economic interests and deeply rooted social dynamics, requiring multidimensional, long-term interventions".

In conclusion, residential segregation and fragmentation of urban space represent major challenges for Moroccan cities, reflecting tensions between the forces of globalization, local urban development dynamics, and aspirations for a more inclusive and cohesive city. Meeting these challenges will require an integrated approach, combining innovative urban planning policies, social cohesion initiatives, and more participatory and inclusive urban governance.

## 2.3.5 Challenges of access to housing and basic urban services

The challenges of access to housing and basic urban services are a central issue in the development of Moroccan cities, reflecting the tensions between rapid urbanization and the authorities' ability to meet the growing needs of the urban population. This issue lies at the heart of the challenges of sustainable urban development and social inclusion in the Moroccan context.

Access to housing remains a major challenge for a significant proportion of Morocco's urban population. Despite the considerable efforts made by the State, notably through social housing programs, supply is struggling to meet growing demand, particularly for low-income households. As Aziz Iraki (2020, p. 87) points out, "the persistence of substandard housing and the development of non-regulated housing bear witness to the limitations of public policies in the face of demographic pressure and land speculation in Morocco's major cities". This situation has led to the proliferation of informal settlements on the outskirts of cities, often lacking essential urban services.

At the same time, access to basic urban services - drinking water, sanitation, electricity, public transport - remains uneven within urban areas. Hind Bouzekri and Mohammed Aderghal (2019, p. 132) note that "disparities in access to basic services between neighborhoods reflect and reinforce socio-spatial inequalities in Moroccan cities". Peri-urban areas and informal settlements are particularly hard hit by these deficiencies, which accentuate their marginalization and limit development opportunities for their inhabitants.

Urban transport is a good illustration of these challenges. Despite major investments in some major cities (tramways in Rabat and Casablanca, for example), public transport provision remains inadequate in the face of urban sprawl and population growth. According to Abderrahmane Rachik (2018, p. 209), "inadequate urban transport systems contribute to reinforcing the spatial and social fragmentation of Moroccan cities, limiting the mobility of the most vulnerable populations".

In response to these challenges, the Moroccan authorities have launched a number of initiatives. The "Cities without Shantytowns" program launched in 2004 has led to significant progress in slum clearance. However, as Lamia Zaki (2021, p. 176) points out, "despite their quantitative success, these programs raise questions about their sustainability and their ability to truly integrate the rehoused populations into the urban and economic fabric".

The challenges of access to housing and basic urban services are also linked to issues of urban governance. Coordination between the various players (State, local authorities, private sector, civil society) remains a major challenge for effective urban planning and management. Mohammed Tozy (2017, p. 98) argues that "improving access to urban services requires an overhaul of local governance modes, favoring greater citizen participation and better cross-sectoral coordination".

In conclusion, meeting the challenges of access to housing and basic urban services in Moroccan cities requires an integrated approach, combining inclusive housing policies, investment in urban infrastructure, and more participatory governance. As Françoise Navez-Bouchanine (2016, p. 243) summarizes, "the challenge for Moroccan cities is to move from a logic of catching up to a forward-looking vision of urban development, integrating social, economic and environmental dimensions in a perspective of sustainability and equity".

## 2.4 Environmental degradation, natural hazards and climate change

### 2.4.1   Air and noise pollution from urbanization

Air and noise pollution linked to urbanization represents a major challenge for Moroccan cities, reflecting the tensions between economic development, rapid urban expansion and the quality of life of city dwellers. These forms of pollution, intrinsically linked to urban growth and human activities, have significant impacts on public health, the environment and the general well-being of urban populations.

Air pollution in Moroccan cities is mainly attributed to three major sources: road traffic, industry, and construction sites. According to a study by Mohammed El Khomri and Fatima Ezzahra Sahli (2019, p. 78), "vehicle emissions, particularly in large conurbations such as Casablanca and Rabat, contribute to more than 60% of urban air pollution, with high concentrations of fine particles (PM2.5 and PM10) and nitrogen oxides (NOx)". This situation is exacerbated by the constant increase in the number of cars on the road and chronic traffic congestion in urban centers.

Although an important source of employment and economic growth, industry also contributes significantly to the degradation of urban air quality. Abdellatif Khattabi and Nadia Machouri (2020, p. 156) point out that "peri-urban industrial zones, often poorly integrated into the urban fabric, generate air pollution that affects adjacent residential neighborhoods, posing considerable health risks for vulnerable populations".

Noise pollution, although less visible, is also a growing issue for urban quality of life in Morocco. The main sources include road traffic, commercial activities and construction sites. As Hassan Radoine (2018, p. 213) notes, "the absence of strict regulations and the lack of urban planning incorporating the acoustic dimension have led to a significant increase in noise levels in Moroccan cities, with negative impacts on the mental and physical health of inhabitants".

Faced with these challenges, Moroccan authorities have begun to introduce policies and regulations aimed at reducing urban pollution. Framework law 99-12 on the environment and sustainable development, adopted in 2014, provided a legal framework for addressing these issues. However, as Fatima-Zahra Taoufik (2021, p. 92) points out, "the effective implementation of these regulations remains a major challenge, due to financial, technical and governance constraints".

Innovative initiatives have also been launched in some cities. For example, the "Green City" project in Benguerir aims to integrate sustainable solutions to reduce urban pollution. According to Mohammed Amine Habba and Khalid Chaoui (2022, p. 178), "this pilot project demonstrates the potential of integrated approaches combining sustainable urban planning, clean technologies and citizen participation to improve urban environmental quality".

Nevertheless, experts stress the need for a more comprehensive and systemic approach to effectively tackle air and noise pollution in Moroccan cities. Fatima El Omari (2020, p. 245) argues that "a transition to more sustainable urban development models requires not only investment in infrastructure and clean technologies, but also a paradigm shift in urban planning and individual behavior".

In conclusion, air and noise pollution linked to urbanization represents a complex and multidimensional challenge for Moroccan cities. Meeting this challenge will require a combination of approaches, including strengthened public policies, investment in green infrastructure, better urban planning, and increased public awareness. As Abdelghani Abouhani (2023, p. 301) summarizes, "improving urban environmental quality in Morocco is not only a public health imperative, but also an essential condition for ensuring the sustainability and attractiveness of Moroccan cities in a context of growing global competition".

## 2.4.2 Waste management and urban sanitation

Waste management and urban sanitation are crucial issues for Moroccan cities, reflecting the complex challenges posed by rapid urbanization, population growth and changing consumption patterns. These issues are at the heart of the environmental and health concerns of Morocco's urban agglomerations, requiring innovative and sustainable solutions.

Solid waste production in Moroccan cities has increased significantly in recent decades. According to Mohammed Ait Hassou and Fatima Ezzahra Ait Brahim (2021, p. 112), "average household waste production in Morocco's urban areas has risen from 0.75 kg/inhabitant/day in 2000 to almost 1 kg/inhabitant/day in 2020, exerting increasing pressure on existing waste management systems". This increase reflects not only demographic growth, but also changing consumption habits linked to urbanization.

Managing these growing volumes of waste poses considerable challenges for Moroccan municipalities. Despite notable progress in waste collection, particularly in large cities, treatment and disposal remain problematic. As Abdellatif El Marjani (2019, p. 87) points out, "the majority of collected waste still ends up in open dumps, with harmful consequences for the environment and public health, including soil and groundwater pollution".

Urban sanitation, closely linked to waste management, represents another major challenge. Although significant progress has been made in extending sanitation networks, many urban areas, particularly peripheral and informal settlements, remain poorly served. Fatima-Zahra Taoufik and Hassan Radoine (2020, p. 156) note that "the inadequacy of sanitation infrastructures in certain urban areas contributes to the pollution of water resources and increases health risks for marginalized populations".

In response to these challenges, Morocco has initiated several programs and reforms aimed at improving waste management and urban sanitation. The National Household Waste Program (PNDM), launched in 2008, has significantly improved waste collection and treatment rates in urban areas. However, as Aziz Iraki (2022, p. 203) notes, "despite the progress made, full implementation of the PNDM faces financial and institutional obstacles, particularly in medium-sized and small towns".

Innovation in waste management is emerging as a promising avenue. Waste sorting and recovery initiatives have been launched in several Moroccan cities. According to Nadia Machouri and Mohammed El Khomri (2023, p. 178), "integrating the informal sector into the recycling value chain represents an opportunity to improve the efficiency of waste management while creating economic opportunities for marginalized populations".

In terms of sanitation, the National Plan for Liquid Sanitation and Wastewater Treatment (PNA) has significantly increased the rate of connection to the sewerage network in urban areas. However, Abdelghani Abouhani (2021, p. 267) points out that "challenges persist in the treatment and reuse of wastewater, requiring significant investment in infrastructure and treatment technologies".

Governance and citizen participation are emerging as key factors in improving waste management and urban sanitation. As Françoise Navez-Bouchanine (2018, p. 312) argues, "a participatory approach involving local communities in the design and implementation of waste management and sanitation solutions can contribute to greater ownership and sustainability of initiatives".

In conclusion, waste management and urban sanitation remain major challenges for Moroccan cities, requiring an integrated approach combining investment in infrastructure, technological innovation, institutional reform and greater citizen involvement. As Mohammed Aderghal (2023, p. 289) summarizes, "improving waste management and sanitation is not only crucial for the environmental and public health of Moroccan cities, but also essential for their sustainable development and resilience in the face of future challenges".

### 2.4.3 Water stress and water supply issues

Water stress and water supply issues are major challenges for Moroccan cities, reflecting the growing tensions between urban demand for water, scarcity of water resources and the impacts of climate change. These issues are at the heart of sustainable development and urban resilience in Morocco.

Morocco, classified as a water-stressed country, is facing increasing pressure on its water resources, particularly in urban areas. According to Mohammed Ait Kadi and Abdelaziz Ziyad (2020, p. 87), "per capita water availability in Morocco has fallen from 2,500 m³ /hab/year in 1960 to less than 650 m³ ZhabZan in 2020, placing the country well below the UN-defined water stress threshold". This situation is exacerbated by rapid urbanization and population growth, which increase pressure on urban water supply systems.

Moroccan cities face multiple challenges when it comes to water supply. On the one hand, urban demand for water is constantly increasing, driven by population growth, rising living standards and economic development. On the other hand, available water resources are increasingly limited and threatened by overexploitation and pollution. As Fatima El Amraoui (2021, p. 156) points out, "the overexploitation of groundwater in peri-urban areas, particularly for intensive agriculture, threatens the long-term sustainability of water supplies in Moroccan cities".

Climate change adds a layer of complexity to these challenges. Climate models predict a decrease in rainfall and an increase in drought episodes in Morocco. According to Abdellatif Khattabi and Nadia Mhammdi (2022, p. 203), "climate projections indicate a potential 10-30% reduction in renewable water resources by 2050, which could exacerbate water tensions in urban areas".

In response to these challenges, Morocco has adopted a multidimensional approach. The national water strategy, launched in 2009, focuses on integrated water resource management, improving the efficiency of distribution networks and developing non-conventional resources. Hassan El Haitami and Fatima Ezzahra Mengoub (2023, p. 178) note that "the deployment of seawater desalination technologies in coastal cities such as Agadir and Casablanca represents a promising alternative for diversifying urban water supply sources".

Improving the efficiency of urban water distribution networks is also a priority. According to Mohammed Rachik and Zineb El Aoufi (2021, p. 245), "losses in urban water distribution networks in Morocco are estimated at between 30% and 40%, underlining the urgent need to invest in the renovation and modernization of urban water infrastructures".

Water demand management is emerging as a crucial area for alleviating urban water stress. Awareness-raising and education initiatives on rational water use have been launched in several Moroccan cities. As Aziz Iraki (2020, p. 312) points out, "the promotion of more sustainable water consumption behaviors and the adoption of water-saving technologies in buildings are essential to reduce pressure on urban water resources".

Urban water governance remains a major challenge. Coordination between the various players (local authorities, water basin agencies, private operators) is crucial to the efficient and equitable management of water resources. Abdelghani Abouhani (2022, p. 289) argues that "improving water governance requires a participatory approach, involving local communities and the private sector in decision-making and the management of urban water resources".

In conclusion, water stress and water supply issues represent complex, multidimensional challenges for Moroccan cities. Meeting these challenges requires a holistic approach, combining investment in infrastructure, technological innovation, institutional reform and changes in consumption behavior. As Françoise Navez-Bouchanine (2023, p. 356) summarizes, "ensuring a sustainable and equitable water supply in Moroccan cities is not only an environmental and health imperative, but also a sine qua non for sustainable urban development and resilience in the face of future challenges".

## 2.4.4 Vulnerability to natural hazards (floods, earthquakes)

The vulnerability of Moroccan cities to natural hazards, particularly floods and earthquakes, is a major challenge for sustainable urban development and resilience. This issue highlights the complex interactions between rapid urbanization, land management and natural hazards in a context of climate change.

Flooding is one of the most recurrent and costly natural hazards for Moroccan cities. According to Mohammed El Jihad and Fatima Ezzahra El Khanchoufi (2021, p. 123), "the increasing waterproofing of soils due to urbanization, combined with the occupation of flood-prone areas and the inadequacy of drainage networks, has significantly increased the vulnerability of urban areas to flooding". The tragic events in Casablanca in 2010 and Tangiers in 2008 illustrate the scale of this threat.

Seismic risk, although less frequent, should not be underestimated. Morocco is located in a convergence zone between the African and Eurasian plates, exposing certain regions to a high seismic risk. As Nadia Mhammdi and Abdelaziz Barakat (2020, p.

187) point out, "rapid and often poorly controlled urbanization, particularly in the north of the country, has led to the construction of many buildings that do not comply with seismic standards, thus increasing the vulnerability of urban populations to this risk".

Faced with these challenges, the Moroccan authorities have undertaken a number of initiatives. The National Flood Control Plan, launched in 2003, has enabled significant progress to be made in protecting urban areas. However, as Abdellatif Khattabi (2022, p. 245) notes, "despite significant investment in protection infrastructure, flood risk management remains mainly reactive rather than proactive, underlining the need for a more integrated approach to urban planning and watershed management".

With regard to seismic risk, Morocco adopted seismic regulations for buildings in 2002 (RPS 2000). However, enforcement remains a challenge, particularly in informal settlements. Fatima- Zahra Taoufik and Hassan Radoine (2023, p. 201) argue that "improving the seismic resilience of Moroccan cities requires not only a strengthening of building standards, but also the regularization and upgrading of existing informal settlements".

Integrating disaster risk reduction into urban planning is emerging as a promising approach. Cities such as Fez and Tangier have begun to draw up development plans incorporating risk mapping. According to Mohammed Aderghal and Hind Bouzekri

(2021, p. 178), "the adoption of a risk-sensitive urban planning approach is essential for reducing the vulnerability of Moroccan cities, but its effective implementation often comes up against financial and institutional constraints".

Climate change adds a further dimension to these challenges. Climate models predict an increase in the frequency and intensity of extreme weather events in Morocco. Aziz Iraki (2022, p. 289) points out that "adapting Moroccan cities to climate change requires a revision of urban planning practices and a systematic integration of resilience considerations into urban development policies".

Citizen participation and risk awareness play a crucial role in building urban resilience. Participatory risk mapping and community-based early warning systems have been tried out in some cities. Françoise Navez-Bouchanine (2020, p. 312) argues that "involving local communities in risk identification and management is essential for developing a culture of prevention and improving the effectiveness of risk reduction measures".

In conclusion, the vulnerability of Moroccan cities to natural hazards, particularly floods and earthquakes, represents a complex challenge requiring a multidimensional approach. As Abdelghani Abouhani (2023, p. 356) summarizes, "strengthening the resilience of Moroccan cities to natural hazards implies not only investment in infrastructure and urban planning, but also a paradigm shift towards proactive and inclusive risk management, integrating the social, economic and environmental dimensions of sustainable urban development".

## 2.4.5 Adapting Moroccan cities to climate change

Adapting Moroccan cities to climate change represents a major, multi-dimensional challenge, highlighting the need for a profound transformation of urban practices in the face of the growing impacts of global warming. This issue lies at the heart of sustainable development and urban resilience in Morocco.

Morocco, by virtue of its geographical position and climatic characteristics, is particularly vulnerable to the effects of climate change. According to Mohammed Ait Kadi and Fatima Ezzahra Mengoub (2021, p. 87), "climate projections for Morocco indicate an increase in average temperatures of 1 to 3.7°C by 2050, accompanied by a drop in precipitation of up to 20%, with profound implications for urban ecosystems". These changes exacerbate existing challenges related to water management, energy and quality of life in Moroccan cities.

Adapting to climate change in the Moroccan urban context requires a holistic, multi-dimensional approach. First and foremost, water resource management is crucial. As Nadia Machouri and Abdellatif Khattabi (2022, p. 156) point out, "adapting urban water supply systems to growing water stress requires not only investment in storage and distribution infrastructures, but also a review of consumption patterns and better integration of rainwater management into urban planning".

Resilience to extreme climatic events is another key area for adaptation. Moroccan cities are increasingly at risk from urban flooding and intense heat waves. Hassan Radoine and Fatima-Zahra Taoufik (2023, p. 203) argue that "adaptation to climate change requires an overhaul of construction and urban planning standards, incorporating nature-based solutions to improve the thermal and hydrological regulation of urban spaces".

Energy efficiency and the transition to renewable energy sources are emerging as key levers for urban adaptation. Morocco has made significant progress in developing solar and wind energy, but integrating them into the urban fabric remains a challenge. According to Aziz Iraki and Mohammed El Khomri (2020, p. 245), "promoting energy efficiency in buildings and developing decentralized energy systems are crucial to reducing the vulnerability of Moroccan cities to energy and climate shocks".

Urban planning plays a central role in adapting to climate change. The systematic integration of climate considerations into urban planning documents and urban development projects is essential. Abdelghani Abouhani (2021, p. 312) points out that "the adoption of a climate-sensitive urban planning approach requires a revision of regulatory and institutional frameworks, as well as increased training in climate issues for urban planning professionals".

Urban mobility is another key area for adaptation. Faced with rising temperatures and the need to reduce greenhouse gas emissions, Moroccan cities are called upon to rethink their transport systems. Françoise Navez-Bouchanine and Mohammed Aderghal (2022, p. 178) note that "the development of efficient public transport systems and the promotion of soft mobility are essential for improving climate resilience and quality of life in Moroccan cities".

Adapting to climate change also raises questions of social equity. The most vulnerable urban populations are often the most exposed to the impacts of climate change. Fatima El Amraoui (2023, p. 289) argues that "climate adaptation in Moroccan cities must integrate a dimension of environmental justice, ensuring that adaptation measures equitably benefit all strata of the urban population".

In conclusion, adapting Moroccan cities to climate change requires a profound transformation of urban practices, involving a multi-sectoral and multi-actor approach. As Mohammed Ait Hassou (2024, p. 356) summarizes, "strengthening the climate resilience of Moroccan cities requires not only investment in infrastructure and technology, but also a paradigm shift in urban governance, favoring innovation, citizen participation and the systematic integration of climate considerations into all urban development decisions". This transition to climate-resilient cities represents both a major challenge and an opportunity to rethink sustainable urban development in Morocco.


## Conclusion:

In conclusion, the second chapter highlights the complex, multi-dimensional challenges facing Moroccan cities in the context of economic globalization and global environmental change. An in-depth analysis of the dynamics of globalization, economic restructuring, demographic pressures and environmental issues reveals an urban landscape undergoing profound change, characterized by tensions between development opportunities and the risks of exacerbating inequalities and vulnerabilities.

Economic globalization, while offering Moroccan cities prospects for growth and international integration, has also spawned processes of economic and social restructuring that have profoundly reconfigured urban spaces. As Mohammed Aderghal and Fatima-Zahra Taoufik (2023, p. 289) point out, "the insertion of Moroccan cities into global economic networks has certainly stimulated the emergence of competitive urban poles, but it has also accentuated socio-spatial disparities and the precariousness of certain segments of the urban population". This duality manifests itself in the coexistence of zones of economic dynamism and enclaves of marginality, reflecting the challenges of inclusiveness and social cohesion in urban development.

Demographic and migratory pressures, combined with the dynamics of economic restructuring, have exacerbated the challenges of access to housing, basic urban services and employment. Rapid and often poorly controlled urban growth has led to the expansion of informal settlements and increased fragmentation of urban space. According to Aziz Iraki and Hind Bouzekri (2022, p. 176), "the resulting residential segregation and urban fragmentation pose major challenges in terms of urban governance and social cohesion, requiring innovative approaches to urban planning and management".

Environmental issues, including pollution, waste management, water stress and adaptation to climate change, are emerging as crucial challenges for the sustainability and resilience of Moroccan cities. These issues underline the urgent need for a transition to more sustainable and resilient urban development models. As Abdellatif Khattabi and Nadia Machouri (2024, p. 312) argue, "adapting Moroccan cities to environmental and climatic challenges requires not only massive investment in green infrastructure and clean technologies, but also a rethinking of urban planning and environmental governance paradigms".

In the face of these multiple, interconnected challenges, the emergence of integrated, multi-sectoral approaches to urban development appears to be a necessity. The transition to more sustainable, inclusive and resilient cities in Morocco implies a reconfiguration of urban governance modes, favoring greater citizen participation, better cross-sectoral coordination and the systematic integration of sustainability and resilience considerations into urban policies and practices.

Ultimately, as Françoise Navez- Bouchanine (2023, p. 356) summarizes, "Moroccan cities are at a critical crossroads, faced with the need to reconcile the imperatives of economic competitiveness, social inclusion and environmental sustainability in a context of globalization and climate change". This situation calls for a profound renewal of urban development approaches, emphasizing innovation, participation and adaptation to emerging challenges. The ability of Moroccan cities to navigate these complex transformations will determine not only their future development trajectory, but also their contribution to achieving sustainable development goals on a national and global scale.

## *Chapter 3: Economic resilience of Moroccan cities in a globalized world*

Against a backdrop of increasing globalization and heightened international competition, the economic resilience of Moroccan cities is a crucial issue for the country's sustainable development. This chapter provides an in-depth analysis of strategies and mechanisms for strengthening the capacity of Moroccan urban centers to adapt and innovate in the face of contemporary economic challenges. It examines how these cities can position themselves as dynamic, competitive economic players, capable of generating inclusive growth and integrating advantageously into global economic networks.

Economic diversification appears to be a fundamental lever for urban resilience. Excessive dependence on a limited number of sectors exposes cities to increased vulnerability to external economic shocks. The identification and development of growth sectors, such as high-tech industries, the digital economy, sustainable tourism and creative industries, are strategic axes for broadening the economic base of Moroccan cities. This diversification must be accompanied by a proactive policy of supporting innovation and strengthening links between academic research and the local economic fabric, in order to foster the emergence of innovative and competitive urban ecosystems.

Investment attractiveness and territorial competitiveness represent a second essential pillar of urban economic resilience. In a world where capital and talent are increasingly mobile, Moroccan cities need to develop their comparative advantages.

to attract and retain investors and skills. This implies a substantial improvement in urban infrastructures, the creation of business-friendly environments, and the implementation of targeted territorial marketing strategies. The development of human capital, through training programs tailored to the needs of the local and international labor markets, is also a key factor in this attractiveness.

Promoting employment, entrepreneurship and the social and solidarity economy is the third major focus of this reflection on urban economic resilience. Faced with the challenges of unemployment and job insecurity, which are particularly acute in urban areas, it is imperative to implement proactive local policies to support job creation and entrepreneurship. The development of start-up-friendly ecosystems, the proliferation of incubators and coworking spaces, and support for social economy initiatives are all levers for stimulating local economic dynamism and promoting the inclusion of vulnerable populations.

Finally, the question of local finances and resource mobilization is a cross-cutting and decisive issue for the effective implementation of urban economic resilience strategies. Strengthening the financial autonomy of local authorities, optimizing local taxation, and seeking out new, innovative financing methods, notably through public-private partnerships and access to green finance, appear to be sine qua non conditions for providing Moroccan cities with the resources they need for sustainable economic development.

The aim of this chapter is to explore these different dimensions of urban economic resilience in depth, based on a rigorous analysis of the issues, opportunities and constraints specific to the Moroccan context. It aims to identify concrete avenues for reflection and action to strengthen the capacity of Moroccan cities to adapt to global economic change and position themselves as dynamic, inclusive growth poles in a constantly evolving world.

## 3.1 Economic diversification and development of new activities

## 3.1.1 Identifying promising economic sectors for Moroccan cities

Identifying promising economic sectors for Moroccan cities is a crucial step in developing sustainable and resilient urban development strategies. This requires an in-depth analysis of the assets and constraints specific to each urban area, as well as a detailed understanding of global economic trends and Morocco's comparative advantages in the international context.

According to a study by Hicham El Moussaoui

and Mohammed Zouiten (2019, p. 78), traditional sectors such as agriculture, textiles and tourism continue to play an important role in the economies of Moroccan cities, but their long-term growth potential is limited by international competition and fluctuations in global markets. The authors underline the need for Moroccan cities to diversify their economic base towards higher value-added sectors with strong innovation potential.

With this in mind, the automotive industry is emerging as a growth sector for several Moroccan cities, notably Tangier and Kénitra. As Abdelkader Kaioua (2020, p. 145) points out, the development of this industry has led to the creation of dynamic industrial ecosystems, fostering the emergence of a fabric of local subcontractors and attracting foreign investment. The author highlights the importance of public policies in structuring these sectors, notably through the creation of free trade zones and the implementation of appropriate training schemes.

The renewable energy sector also represents a promising area of development for many Moroccan cities. Fatima Arib (2021, p. 203) highlights Morocco's considerable potential in solar and wind energy, and emphasizes the opportunity for cities to position themselves as poles of excellence in this field. The author stresses the need for an integrated approach, combining research, training and industrial development, to maximize the local economic benefits of these investments.

The digital economy and information and communication technologies (ICT) are emerging as another key sector for the economic development of Moroccan cities. Rachid El Houdaigui and Nabil Adel (2018, p. 92) analyze Casablanca's emergence as a regional hub for financial services and ICT, highlighting the catalytic role of investment in digital infrastructure and the training of local talent. The authors highlight the importance of creating innovation-friendly ecosystems, including incubators, gas pedals and venture capital funds, to stimulate the creation of start-ups and attract international technology companies.

The pharmaceutical and biotech industry also appears to be a diversification opportunity for certain Moroccan cities. Nadia Benabdeljlil and Youssef Toumi (2020, p. 167) examine the development potential of this sector, particularly in Rabat and Casablanca, drawing on existing expertise in the production of generic drugs and investment in research and development. The authors stress the importance of close collaboration between universities, research centers and industry to foster innovation and technology transfer.

Finally, the green and circular economy is emerging as a promising cross-cutting sector for Moroccan cities. Mohamed Behnassi and Meryem El Alaoui (2022, p. 211) analyze the opportunities offered by the development of recycling, waste recovery and eco-design industries in urban areas. The authors highlight the potential for local job creation and improved urban quality of life associated with these activities, while underlining the need for an appropriate regulatory and incentive framework to stimulate their development.

In conclusion, identifying promising economic sectors for Moroccan cities requires a multidimensional approach, taking into account local specificities, global trends and sustainable development objectives. Diversification into high value-added sectors, innovation and the creation of integrated economic ecosystems appear to be essential levers for strengthening the economic resilience of Moroccan cities in the face of the challenges of globalization.

## 3.1.2 Development of high-tech industries and the digital economy

The development of high-tech industries and the digital economy represents a major strategic challenge for Moroccan cities, as part of a dynamic transition towards a knowledge and innovation-based economy. This is crucial to boosting the international competitiveness of Moroccan urban areas and creating skilled jobs, thus meeting the aspirations of a young and increasingly educated population.

According to Mohammed Bougroum and Aomar Ibourk (2019, p. 156), Morocco has made significant progress in developing its digital infrastructure over the past decade, creating a favorable foundation for the growth of high-tech industries. In particular, the authors highlight the importance of investments in high-speed telecoms networks and the creation of specialized technopoles. However, they warn of the risk of a territorial digital divide, calling for a balanced digital development policy across the country's different regions.

Casablanca's emergence as a regional technology hub illustrates the development potential of the digital economy in Moroccan cities. Nabil El Mabrouki and Karim Hasnaoui (2020, p. 213) analyze the success factors of this dynamic, highlighting the crucial role of public-private partnerships in the creation of innovation ecosystems. The authors stress the importance of setting up support structures such as incubators and gas pedals, as well as access to financing for technology start-ups.

The development of high-tech industries in Moroccan cities is also based on a strategy of intelligent specialization. Fatima Zahra Sossi Alaoui and Hassan Zaoual (2021, p. 178) examine the case of Tangier and its positioning in the aeronautics industry. Their study highlights the importance of training a skilled workforce and creating synergies between companies, universities and research centers to foster innovation and technology transfer.

The rise of the digital economy in Moroccan cities is also reflected in the rapid development of the digital services sector. Rachid Aboulaich and Moulay Driss El Mellouki (2022, p. 245) analyze the emergence of new forms of employment linked to the platform economy and remote working. The authors highlight the opportunities offered by these developments in terms of flexibility and access to employment, while warning against the risks of casualization and calling for appropriate regulation of these new forms of work.

Cybersecurity is emerging as a crucial area for the development of high-tech industries and the digital economy in Moroccan cities. Youssef Benkirane and Abdelhamid Benmoussa (2020, p. 301) emphasize the strategic importance of this sector, both for the protection of critical infrastructures and for the development of exportable national expertise. The authors recommend the introduction of specialized training courses and the creation of cybersecurity centers of excellence in the country's main university towns.

Artificial intelligence (AI) and big data are also emerging as promising fields for Moroccan cities. Salma Lalaoui and Mohammed Essaaidi (2021, p. 189) analyze the potential of these technologies to optimize urban management and develop innovative services for citizens. However, the authors stress the need for a suitable ethical and regulatory framework to govern the use of these technologies and protect citizens' personal data.

In conclusion, the development of high-tech industries and the digital economy in Moroccan cities requires a holistic approach, combining investment in infrastructure, human capital training, support for innovation and the creation of a favorable regulatory environment. As Nadia El Ghazouani and Driss Khrouz (2022, p. 267) point out, this transition to a knowledge-based economy is essential to ensure the long-term competitiveness of Moroccan cities and create skilled employment opportunities for future generations. Nevertheless, the authors stress the importance of an inclusive approach, ensuring that the benefits of this digital transition are equitably distributed across urban society.

### 3.1.3 Promoting sustainable tourism and enhancing cultural heritage

The promotion of sustainable tourism and the enhancement of cultural heritage are emerging as major strategic axes for the economic development of Moroccan cities, combining the preservation of local identity with international appeal. This approach aims to reconcile the imperatives of economic growth with the principles of environmental sustainability and respect for local communities.

According to Hassan Ramou and Mohamed Aderghal (2019, p. 123), sustainable tourism offers Moroccan cities an opportunity to differentiate themselves on the international tourism market, by highlighting their authenticity and cultural richness. The authors stress the importance of an integrated approach, combining heritage preservation, environmentally-friendly infrastructure development and the involvement of local communities in tourism management. They warn against the risk of museumizing historic centers, and advocate a dynamic vision of heritage, integrating living cultural practices.

The valorization of intangible cultural heritage appears to be an important lever for the development of sustainable tourism in Moroccan cities. Fatima Bouchmal and Ahmed Skounti (2020, p. 178) analyze the potential of traditional know-how, festivals and culinary practices as tourist attractions. Their study highlights the importance of training local players and establishing quality labels to guarantee the authenticity of the experiences offered to visitors.

The integration of new technologies in the promotion of sustainable tourism and the enhancement of cultural heritage offers promising prospects. Nadia Mzoughi and Salma Ait Taleb (2021, p. 245) examine the use of augmented reality and mobile applications to enrich the visitor experience at urban historic sites. The authors highlight the potential of these tools to attract a younger audience and to disseminate information on best practices in responsible tourism.

Managing tourist flows in Morocco's historic cities is a major challenge for heritage preservation and residents' quality of life. Mohammed El Faiz and Ouidad Tebbaa (2018, p. 201) analyze the strategies implemented in Marrakech to reconcile tourist attractiveness and preservation of the medina. The authors advocate an integrated management approach, combining flow regulation, diversification of the tourism offer and visitor awareness of preservation issues.

The development of urban ecotourism is emerging as a promising trend for Moroccan cities. Brahim El Fasskaoui and Mohamed Kadiri (2022, p. 167) explore the potential of urban and peri-urban green spaces as sustainable tourist attractions. Their study highlights the opportunities for green job creation and environmental awareness offered by these initiatives, while emphasizing the need for careful planning to avoid negative impacts on fragile ecosystems.

The training and professionalization of tourism stakeholders are crucial to the development of high-quality sustainable tourism. Rachida Saigh Bousta and Mohamed Berriane (2020, p. 289) analyze skills needs in the sustainable tourism sector and suggest ways of adapting training curricula to the challenges of sustainability. The authors stress the importance of developing cross-disciplinary skills, combining cultural knowledge, environmental awareness and mastery of digital tools.

Participatory governance of tourism and heritage is emerging as a key success factor for sustainable development. Amina Hachimi Alaoui and Hassan Zaoual (2021, p. 312) examine experiences of participatory heritage management in several Moroccan cities. Their study highlights the importance of involving local communities in the definition of tourism and heritage strategies, in order to ensure a fair distribution of benefits and preserve the authenticity of places.

In conclusion, promoting sustainable tourism and enhancing cultural heritage in Moroccan cities requires a multi-dimensional approach, combining preservation, innovation and citizen participation. As Mimoun Hillali and Said Boujrouf (2022, p. 335) point out, this approach offers Moroccan cities the opportunity to position themselves as prime destinations for responsible, culturally enriching tourism. Nevertheless, the authors stress the need for constant vigilance to maintain the balance between tourism development and preservation of local identity, in a context of increasing pressure on urban and heritage resources.

### 3.1.4 Booming creative and cultural industries

The rise of the creative and cultural industries (CCI) in Moroccan cities represents an innovative and meaningful vector for economic development, combining the enhancement of cultural heritage and contemporary creation. This sector, at the crossroads of art, technology and entrepreneurship, offers promising prospects for the economic diversification and international influence of Moroccan urban centers.

According to an in-depth study by Driss Ksikes and Kenza Sefrioui (2019, p. 145), CCIs in Morocco are experiencing significant growth, particularly in the fields of audiovisual, design, fashion and digital arts. The authors highlight the potential of these industries for creating skilled jobs and stimulating urban innovation. However, they warn of regional disparities in the development of these sectors, calling for a coherent national policy to support the emergence of creative ecosystems across Morocco's urban territory.

Casablanca is emerging as a major hub for CCI in Morocco. Fatima Zahra El Malki and Hassan Zouaoui (2020, p. 212) analyze the factors that have contributed to this dynamic, highlighting the role of public and private initiatives in the creation of spaces dedicated to artistic creation and exhibition. The authors stress the importance of rehabilitating industrial heritage into cultural spaces, as a catalyst for urban regeneration and creative appeal.

Morocco's film industry is undergoing remarkable development, making a significant contribution to the economy of its cities. Jamal Eddine Naji and Noureddine Bensalah (2021, p. 178) examine the economic and cultural impact of Ouarzazate's film studios. Their study highlights the positive spin-offs in terms of local employment and film tourism, while underlining the challenges of training local talent and diversifying production beyond international shoots.

The fashion and design sector is emerging as an important pillar of CCI in Moroccan cities. Meriem El Bouhali and Rachid Zenati (2020, p. 267) analyze the development potential of this sector, using the example of Marrakech and its International Fashion Festival. The authors highlight the importance of fusing traditional know-how with contemporary creativity, as well as the role of design schools in nurturing local talent capable of competing on the international scene.

Digital arts and new media represent a rapidly expanding field within Moroccan CCIs. Younès Baba-Ali and Dounia Benslimane (2022, p. 201) explore emerging initiatives in this field, particularly in Rabat and Tangier. Their research highlights the potential of these artistic practices to attract a young, connected audience, while also highlighting the challenges associated with technological infrastructure and specialized training.

As a creative industry, music plays a crucial role in the cultural economy of Moroccan cities. Ahmed Aydoun and Amine Hamma (2021, p. 289) analyze the economic impact of music festivals such as the Mawazine Festival in Rabat and the Gnaoua Festival in Essaouira. The authors highlight the positive spin-offs in terms of cultural tourism and seasonal job creation, while stressing the importance of supporting the local music scene throughout the year.

Art crafts, at the frontier between tradition and contemporary creation, constitute a key CCI sector in Morocco. Fatima Sadiqi and Moha Ennaji (2020, p. 234) examine initiatives to modernize and promote Moroccan crafts, particularly in Fez and Tetouan. Their study highlights the importance of ongoing training for artisans and the creation of quality labels to position Moroccan crafts on international luxury and design markets.

In conclusion, the rise of the creative and cultural industries in Moroccan cities represents an important lever for economic development and international influence. As Mohammed Tozy and Youssef Courbage (2022, p. 312) point out, this sector offers unique opportunities to enhance Morocco's cultural wealth while stimulating innovation and urban creativity. Nevertheless, the authors insist on the need for a coherent policy to support these industries, including appropriate funding mechanisms, stronger intellectual property protection and an ambitious international promotion strategy. The challenge lies in the ability of Moroccan cities to create dynamic creative ecosystems, capable of retaining local talent and attracting international creators, while preserving the authenticity and cultural diversity that make Morocco so rich.

### 3.1.5 Strengthening links between research, innovation and the local economic fabric

Strengthening the links between research, innovation and the local economic fabric is crucial to the sustainable development and competitiveness of Moroccan cities in a globalized economic context. This synergy between the academic world, research centers and local businesses is essential to stimulate innovation, promote technology transfer and create a dynamic for endogenous growth.

According to an in-depth study by Abdellatif Komat and Nadia Benabdeljlil (2019, p. 178), Morocco has made significant progress in research and development (R&D) over the past decade, but transferring this knowledge to the local economic fabric remains a major challenge. The authors stress the importance of creating effective interface structures between universities and businesses, such as technology transfer offices and university incubators. They highlight the crucial role of public policy in creating an environment conducive to such collaboration.

The creation of competitiveness clusters in Moroccan cities appears to be a promising strategy for strengthening links between research and industry. Mohammed Bensaid and Fatima Zahra Aziz (2020, p. 245) analyze the experience of the Casablanca Technopark and its impact on the local innovation ecosystem. Their study highlights the importance of geographical proximity between research, innovation and industry players in fostering knowledge exchange and stimulating the creation of innovative start-ups.

Adapting university curricula to the needs of the local economic fabric is a major challenge if this synergy is to be strengthened. Rajaa Cherkaoui El Moursli and Hassan Sahbi (2021, p. 312) examine initiatives aimed at bringing academia closer to the realities of the job market. The authors stress the importance of in-company internships, collaborative research projects and the involvement of professionals in curriculum development to produce graduates capable of meeting the innovation needs of local businesses.

The development of applied research geared to the specific needs of the local economic fabric is a priority. Khalid Soudi and Asmae Diani (2020, p. 189) analyze collaborative research initiatives in the field of sustainable agriculture and agrifood in Meknes. Their study highlights the positive impact of these collaborations on innovation in local SMEs and on improving the competitiveness of the regional agricultural sector.

The mobility of researchers between academia and industry appears to be an important vector for knowledge transfer. Nouzha Boujettou and Mohammed Youssfi (2022, p. 267) examine the mechanisms put in place to encourage this mobility, such as company creation leave or associate researcher positions in companies. The authors stress the importance of these mechanisms in fostering mutual understanding between the two worlds and stimulating innovation.

Financing innovation and collaborative research is a crucial factor in strengthening the links between research and the local economic fabric. Rachid Guerraoui and Ghita Lahlou (2021, p. 201) analyze the various financing instruments available in Morocco, such as seed funds, collaborative research grants and R&D tax incentives. Their study highlights the need for a differentiated approach depending on the sector and stage of development of innovation projects.

The protection and valorization of intellectual

property are emerging as cross-cutting issues in stimulating collaboration between research and industry. Salwa Belkeziz and Omar Ouhejjou (2020, p. 289) examine the challenges of patenting innovations from public research and revenue sharing between researchers and institutions. The authors stress the importance of a clear, incentive-based legal framework to encourage researchers to valorize their discoveries and collaborate with the private sector.

In conclusion, strengthening the links between research, innovation and the local economic fabric in Moroccan cities requires a systemic and multidimensional approach. As Driss Khrouz and Nadia El Ghazouani (2022, p. 335) point out, this synergy is essential to create dynamic innovation ecosystems, capable of generating endogenous growth and positioning Moroccan cities as innovation hubs on a regional and international scale. To realize the full potential of this collaboration between research and industry, the authors stress the need for a long-term vision, sustained commitment from public authorities and a culture of innovation shared by all players. The challenge lies in the ability of Moroccan cities to create an environment conducive to open innovation, where the boundaries between fundamental research, applied research and industrial development are blurred in favor of a collective dynamic that creates value and resolves societal challenges.

## 3.2 Investment attractiveness and territorial competitiveness

### 3.2.1 Improving urban infrastructure and connectivity

Improving urban infrastructure and connectivity is a fundamental pillar for the economic development and competitiveness of Moroccan cities in a globalized world. Infrastructure modernization is essential to attracting investment, improving citizens' quality of life and positioning cities as key players in the regional and international economy.

According to an in-depth analysis by Mohammed Berrada and Driss Guerraoui (2019, p. 156), Morocco's massive investments in urban infrastructure over the past two decades have significantly improved the connectivity and attractiveness of the country's main cities. In particular, the authors highlight the positive impact of major urban transport projects, such as the Casablanca and Rabat tramways, on urban mobility and the accessibility of economic activity zones. However, they warn of the risk of an infrastructure divide between major metropolises and medium-sized towns, and call for a more balanced regional planning policy.

The modernization of telecommunications infrastructures is becoming a crucial issue for the economic attractiveness of Moroccan cities. Nadia Mansouri and Hassan Haddouch (2020, p. 213) examine the deployment of fiber optic networks and 5G in Morocco's urban areas. Their study highlights the importance of these digital infrastructures for the development of the digital economy, the attraction of technology companies and the implementation of "smart city" solutions. However, the authors stress the need for appropriate regulation to ensure healthy competition and fair coverage of urban areas.

Improving intercity transport infrastructure plays a key role in strengthening the connectivity of Moroccan cities. Abdellatif Chahid and Fatima Arib (2021, p. 278) analyze the impact of the high-speed rail network on the economic dynamics of the cities it serves. Their research highlights the positive effects in terms of attractiveness to investors and mobility of skilled labor, while underlining the importance of properly integrating stations into the urban fabric to maximize local economic spin-offs.

Sustainable management of water resources and wastewater infrastructure is a major challenge for Moroccan cities. Mohammed Ait Kadi and Nabil Ben Khatra (2020, p. 189) examine the strategies put in place to improve the efficiency of drinking water and wastewater treatment networks. The authors emphasize the importance of these infrastructures for public health, urban quality of life and economic attractiveness, while highlighting the potential of the circular economy in urban water resource management.

The development of sustainable energy infrastructures appears to be a strategic axis for the competitiveness of Moroccan cities. Amal Mecherfi and Said Mouline (2022, p. 245) analyze the deployment of renewable energies in urban environments, particularly through solar roof and smart microgrid projects. Their study highlights the potential of these infrastructures to reduce cities' energy dependency, create local jobs and attract investment in green technologies.

Improving urban logistics infrastructures is becoming a crucial factor in making economic exchanges more fluid. Rachid Saikouk and Laila Ouhajjou (2021, p. 301) examine initiatives to optimize last-mile logistics in major Moroccan cities, notably through the creation of urban distribution centers and the use of electric vehicles. The authors emphasize the importance of these infrastructures in reducing urban congestion and improving the efficiency of local supply chains.

The resilience of urban infrastructures to climate risks is emerging as a major concern. Hassan Radoine and Fatima-Zahra Kounssi (2020, p. 167) analyze strategies for adapting Moroccan urban infrastructures to the challenges of climate change. Their study highlights the importance of integrating climate resilience into infrastructure planning and design to ensure the continuity of urban services and the long-term security of investments.

In conclusion, improving urban infrastructure and connectivity in Moroccan cities requires an integrated and visionary approach. As Mohammed Tawfik Mouline and Laila Hilal (2022, p. 335) point out, this modernization is essential to position

Moroccan cities as competitive economic hubs on a regional and international scale. The authors stress the need for long-term strategic planning, effective coordination between different levels of governance and an innovative approach to infrastructure financing, notably through public-private partnerships and green finance mechanisms. The challenge lies in the ability of Moroccan cities to develop smart, sustainable and inclusive infrastructures, capable of supporting balanced economic growth while meeting the environmental and social challenges of the 21st century.

### 3.2.2 Creation of special economic zones and competitiveness clusters

The creation of special economic zones (SEZs) and competitiveness clusters is a key strategy for boosting the economic development and attractiveness of Moroccan cities in a context of heightened international competition. These initiatives aim to concentrate resources, skills and investment in defined geographical areas, thereby fostering the emergence of high-performance industrial and innovative ecosystems.

According to an in-depth study by Abdelkader Kaioua and Nouzha Chekrouni (2019, p. 178), SEZs in Morocco have played a crucial role in attracting foreign direct investment and diversifying the country's industrial fabric. In particular, the authors analyze the success of the Tangier Med free zone, which has enabled the emergence of a world-class automotive cluster. However, they emphasize the importance of further integrating these zones with the local economic fabric to maximize spin-offs in terms of employment and technology transfer.

The design and governance of competitiveness clusters represent a major challenge to their effectiveness. Fatima Zahra El Malki and Hassan Zaoual (2020, p. 245) examine the success factors and challenges faced by Moroccan technopoles, such as Casanearshore or Rabat Technopolis. Their research highlights the importance of participative governance, involving public, private and academic players, as well as the need for a clear, shared strategic vision to guide the development of these clusters.

Adapting SEZs and competitiveness clusters to regional specificities appears to be a key success factor. Mohammed Refass and Driss Khrouz (2021, p. 312) analyze smart specialization strategies implemented in different regions of Morocco. The authors stress the importance of capitalizing on local assets and know-how to create unique, internationally competitive economic ecosystems.

The development of research and innovation infrastructures within SEZs and competitiveness clusters is a crucial issue. Rachid El Houdaigui and Nabil Adel (2020, p. 201) examine the role of R&D centers and shared technology platforms in stimulating innovation within these areas. Their study highlights the importance of these infrastructures in attracting innovative companies and fostering collaboration between public research and industry.

Training and human capital development appear to be priorities for the success of SEZs and competitiveness clusters. Amina Benkhadra and Said Benhajjou (2022, p. 289) analyze the vocational training and higher education initiatives set up in partnership with economic players within these zones. The authors emphasize the importance of a close match between the training on offer and the needs of companies, to guarantee the employability of graduates and the attractiveness of the zones to investors.

The integration of sustainability issues into the design and operation of SEZs and competitiveness clusters is emerging as a strong trend. Fatima Arib and Mohammed Behnassi (2021, p. 234) examine initiatives to create "industrial eco-parks" in Morocco, incorporating circular economy and energy efficiency principles. Their research highlights the potential of these approaches to reduce the environmental footprint of industrial activities and create new economic opportunities in green sectors.

Financing and tax incentives play a key role in the attractiveness of SEZs and competitiveness clusters. Najib Akesbi and Omar El Hyani (2020, p. 267) analyze the various incentive mechanisms put in place in Morocco to attract investors to these zones. The authors stress the importance of striking a balance between fiscal attractiveness and contribution to local development, calling for reflection on the long-term sustainability of these incentive models.

In conclusion, the creation of special economic zones and competitiveness clusters in Moroccan cities is part of an overall strategy to strengthen national economic competitiveness. As Nadia El Ghazouani and Larabi Jaidi (2022, p. 335) point out, these initiatives offer unique opportunities to accelerate industrial modernization, stimulate innovation and create skilled jobs. However, the authors stress the need for an integrated approach, taking into account the economic, social and environmental dimensions of urban development. The challenge lies in the ability of Moroccan cities to create dynamic and inclusive economic ecosystems, capable of integrating advantageously into global value chains while generating positive spin-offs for local development. The success of these initiatives will depend on the ability to maintain a long-term strategic vision, to ensure effective and participative governance, and to continually adapt these areas to the rapid changes taking place in the global economy.

## 3.2.3 Simplification of administrative procedures and business climate

Simplifying administrative procedures and improving the business climate are crucial to boosting the attractiveness and competitiveness of Moroccan cities in a globalized economic context. These reforms are aimed at reducing bureaucratic obstacles, speeding up the process of setting up and developing businesses, and creating an environment conducive to investment and innovation.

According to an in-depth study by Nadia Salah and Mohamed Berrada (2019, p. 156), Morocco has made significant progress in improving its business climate over the past decade, thanks in particular to the introduction of one-stop shops and the dematerialization of certain administrative procedures. The authors point out, however, that disparities persist between the kingdom's different cities in terms of administrative efficiency, calling for nationwide harmonization of practices to guarantee territorial equity in access to economic opportunities.

The digitization of administrative services is emerging as a major lever for simplification and efficiency. Rachid Guerraoui and Ghita Lahlou (2020, p. 213) analyze the impact of e-government on facilitating administrative procedures for businesses in Morocco's main cities. Their study highlights the time and cost savings for entrepreneurs, while stressing the importance of supporting this digital transition with training and awareness-raising measures, particularly for small and medium-sized businesses.

Reform of the legal and regulatory framework for business appears to be a key element in improving the business climate. Omar Aloui and Fatima-Zohra Alaoui (2021, p. 278) examine recent amendments to

Morocco's Commercial Code and Company Law, aimed at facilitating business start-ups and strengthening investor protection. The authors stress the importance of these reforms in bringing the Moroccan legal framework into line with international standards, while calling for effective and uniform application of these new provisions at local level.

Improving access to land and building permits is a major challenge for business development in Moroccan cities. Mohammed Tozy and Beatrice Hibou (2020, p. 189) analyze the initiatives put in place to simplify procedures for allocating industrial land and obtaining building permits. Their research highlights the importance of transparency and fairness in these processes to prevent corruption and ensure efficient allocation of land resources.

Local tax reform is an important lever for improving the urban business climate. Najib Akesbi and Mohammed Haddy (2022, p. 245) examine measures to simplify and rationalize local taxation in Morocco. The authors stress the importance of striking a balance between tax attractiveness for investors and the need to maintain sufficient resources for local development, proposing avenues for a tax system that is more incentive-based and better adapted to local economic realities.

Improving the efficiency of the judicial system in commercial matters appears to be a key confidence-building factor for investors. Nadia Bernoussi and

Ahmed Ouazzani Chahdi (2021, p. 301) analyze reforms aimed at strengthening the independence and competence of commercial courts in Morocco. Their study highlights the importance of ongoing training for magistrates and the modernization of judicial procedures to speed up the settlement of commercial disputes and enhance the legal security of transactions.

The fight against corruption and the promotion of integrity in economic transactions are priorities for improving the business climate. Transparency Maroc and Azeddine Akesbi (2020, p. 167) examine local initiatives to promote transparency and ethics in relations between companies and administrations. The authors stress the importance of effective control mechanisms and a culture of integrity in restoring the confidence of investors and citizens in public institutions.

In conclusion, simplifying administrative procedures and improving the business climate in Moroccan cities requires a holistic and coordinated approach. As emphasized by Driss Guerraoui and Nadia El Ghazouani (2022, p. 335), these reforms are essential to unleash entrepreneurial potential, attract investment and stimulate job creation in urban areas. The authors stress the need for strong political will, effective coordination between different levels of governance, and active involvement of the private sector and civil society in the design and implementation of these reforms. The challenge lies in the ability of Moroccan cities to create an agile, transparent and predictable administrative and regulatory environment, capable of adapting rapidly to changes in the global economy while preserving the general interest and local specificities. The success of these initiatives to simplify and improve the business climate will be crucial in positioning Moroccan cities as attractive destinations for national and international investors, and in stimulating sustainable and inclusive economic growth on an urban scale.

### 3.2.4 Territorial marketing strategies and international promotion

Territorial marketing and international promotion strategies are essential levers for boosting the attractiveness and competitiveness of Moroccan cities on the global economic stage. These approaches aim to build and promote an attractive, differentiated image of urban areas, in order to attract international investors, talent and visitors.

According to an in-depth analysis by Hassan Zaoual and Fatima Zahra El Malki (2019, p. 178), territorial marketing in Morocco has undergone a significant evolution, moving from an essentially tourism-based approach to a more comprehensive economic promotion strategy. The authors stress the importance of an integrated approach, articulating cities' economic, cultural and environmental assets to build a unique and attractive value proposition. However, they warn against the risk of standardizing promotional strategies, calling for a genuine enhancement of local specificities.

Building a strong territorial brand appears to be a key element of urban marketing. Nouzha Chekrouni and Abdelkader Kaioua (2020, p. 213) examine the "city branding" initiatives implemented by several major Moroccan cities. Their study highlights the importance of a participatory approach, involving all local stakeholders in defining the identity and values of the city brand. The authors also emphasize the need for consistency between the projected image and the lived reality of the territory to guarantee the credibility and sustainability of the branding strategy.

The use of digital tools and social networks is becoming a major focus of Moroccan cities' international promotion strategies. Driss Ksikes and Kenza Sefrioui (2021, p. 245) analyze the impact of digital communication campaigns run by several Moroccan metropolises to attract international investors and talent. Their research highlights the potential of these tools to target specific audiences and measure the effectiveness of promotional actions, while underlining the importance of a content strategy adapted to different platforms and cultures.

Participation in international events and networks is emerging as an important vector for the visibility and positioning of Moroccan cities. Mohammed Berriane and Bouchta El Moumni (2020, p. 301) examine the impact of organizing major international events, such as COP22 in Marrakech, on the image and attractiveness of host cities. The authors stress the importance of a legacy strategy to capitalize on these events and create lasting benefits for the territory, particularly in terms of infrastructure and skills.

City economic diplomacy is becoming a strategic axis of international promotion. Rachid El Houdaigui and Nabil Adel (2022, p. 189) analyze the paradiplomacy initiatives undertaken by several major Moroccan cities to forge international economic partnerships. Their study highlights the importance of these cooperation networks in facilitating access to foreign markets, attracting investment and promoting the exchange of best practices in urban development.

Enhancing the value of cultural heritage and urban creativity appears to be a major lever for differentiation and international appeal. Fatima Sadiqi and Moha Ennaji (2021, p. 267) examine promotional strategies based on the creative economy and cultural tourism in several Moroccan cities. The authors emphasize the importance of preserving the authenticity of heritage while incorporating it into a dynamic of modernity and innovation, in order to create a unique and attractive cultural offer for visitors and international investors.

The evaluation and benchmarking of territorial marketing strategies are crucial to the continuous improvement of practices. Nadia El Ghazouani and Larabi Jaidi (2020, p. 234) analyze the methodologies and indicators used to measure the effectiveness of

Moroccan cities' international promotion initiatives. Their research highlights the importance of a multidimensional approach, integrating quantitative and qualitative criteria, to assess the real impact of territorial marketing strategies on the attractiveness and economic development of cities.

In conclusion, territorial marketing and international promotion strategies for Moroccan cities are part of a global approach to strategic positioning in the global economy. As Mohammed Tozy and Beatrice Hibou (2022, p. 335) point out, these approaches require a long-term vision, effective coordination between public and private players, and the ability to constantly adapt to changes in the perceptions and expectations of international target audiences. The authors stress the importance of developing authentic, differentiated strategies that are rooted in local realities while meeting international standards of attractiveness. The challenge lies in the ability of Moroccan cities to build and project a coherent, attractive image, capable of distinguishing them in a context of increased territorial competition, while generating tangible benefits for local economic and social development. The success of these marketing and international promotion initiatives will be crucial in positioning Moroccan cities as destinations of choice for international investors, talents and visitors, thus contributing to their influence and prosperity on a global scale.

### 3.2.5 Human capital development and appropriate vocational training

The development of human capital and the introduction of appropriate vocational training are crucial to boosting the competitiveness and attractiveness of Moroccan cities in a rapidly changing economic context. These strategic axes aim to align the skills of the local workforce with the evolving needs of the labor market, and to stimulate innovation and productivity in urban economic ecosystems.

According to an in-depth study by Driss Khrouz and Noureddine El Aoufi (2019, p. 167), Morocco has made significant progress in extending access to education and vocational training, but challenges persist in terms of matching the skills developed with the needs of the urban labor market. The authors stress the importance of a forward-looking approach to planning training programs, taking into account long-term technological trends and sectoral mutations.

Involving the private sector in the design and implementation of vocational training programs appears to be a key success factor. Fatima Zahra Alaoui and Omar Aloui (2020, p. 213) analyze the experience of public-private partnerships in vocational training in Morocco. Their study highlights the positive impact of these collaborations on the employability of graduates and on matching skills to the specific needs of local businesses. However, the authors stress the need for a clear regulatory framework to encourage sustainable private-sector involvement in these initiatives.

The development of cross-disciplinary competencies and soft skills is emerging as a priority to prepare the urban workforce for the challenges of the 21st century economy. Nadia Benabdejlil and Saïd Hanchane (2021, p. 278) examine initiatives to integrate these skills into initial and continuing training curricula in Morocco. Their research highlights the importance of developing adaptability, creativity and collaborative skills to meet the demands of an ever-changing job market.

Lifelong learning and professional retraining are emerging as crucial issues in the face of rapid changes in the urban economy. Mohammed Dahbi and Rachida Saigh Bousta (2020, p. 301) analyze the mechanisms in place to facilitate lifelong learning and professional mobility in Morocco's main cities. The authors emphasize the importance of developing a culture of lifelong learning, and of setting up mechanisms for the recognition and validation of prior experience to promote the flexibility and adaptability of the urban workforce.

The integration of digital technologies into training processes is emerging as a major lever for improving the accessibility and effectiveness of educational programs. Abdellatif Miraoui and Khalid Soudi (2022, p. 245) examine the potential of MOOCs (Massive Open Online Courses) and e-learning platforms to democratize access to vocational training in Moroccan cities. Their study highlights the opportunities offered by these tools to reach a wider audience and offer personalized training paths, while underlining the importance of appropriate support to guarantee the effectiveness of these distance learning modalities.

The development of entrepreneurship and innovation is emerging as a cross-cutting theme in the development of urban human capital. Fatima Zahra Zaouli and Hassan Zaoual (2021, p. 189) analyze initiatives to promote entrepreneurial culture and innovation skills in Moroccan training curricula. The authors highlight the importance of these approaches in stimulating job creation and the emergence of innovative start-ups in urban economic ecosystems.

Inclusion and diversity in vocational training programs appear to be crucial issues for equitable economic development in Moroccan cities. Aïcha Belarbi and Mohammed Bensaid (2020, p. 234) examine the strategies put in place to promote access to skills training for women and marginalized groups. Their research highlights the importance of these initiatives in reducing socio-economic inequalities and fully mobilizing the human potential of urban areas.

In conclusion, the development of human capital and the introduction of appropriate vocational training are at the heart of strategies to boost the competitiveness and attractiveness of Moroccan cities. As Nadia El Ghazouani and Driss Guerraoui (2022, p.

335) point out, these approaches require a systemic and forward-looking vision, integrating the economic, technological and social dimensions of urban development. The authors stress the importance of close collaboration between public, private and academic players to create agile and innovative training ecosystems, capable of adapting rapidly to changes in the global economy. The challenge lies in the ability of Moroccan cities to develop a highly skilled and adaptable workforce, capable of supporting growth in high value-added sectors while promoting social inclusion and innovation. The success of these human capital development initiatives will be crucial in positioning Moroccan cities as poles of attraction for international investors and talent, thus contributing to their economic resilience and their influence on a regional and global scale.

## 3.3 Employment, entrepreneurship and the social economy

### 3.3.1 Local policies to support job creation

Local policies to support job creation are a major strategic focus for the economic and social development of Moroccan cities. These initiatives aim to stimulate local job growth, reduce unemployment and promote economic inclusion in an urban context marked by major demographic and socio-economic challenges.

According to an in-depth study by Noureddine El Aoufi and Mohammed Bensaïd (2019, p. 156),

Moroccan cities are facing increasing pressure on the labor market, particularly due to the continued influx of rural populations and the massive arrival of young graduates. The authors stress the importance of a multidimensional, territorialized approach to employment policies, taking into account the economic and social specificities of each urban area.

Support for small and medium-sized enterprises (SMEs) appears to be an essential lever in local job creation policies. Fatima Zahra Alaoui and Omar Aloui (2020, p. 213) analyze the impact of SME support and financing programs implemented in several Moroccan cities. Their study highlights the multiplier effect of these initiatives on direct and indirect job creation, while underlining the importance of an integrated support ecosystem, including training, mentoring and access to markets.

The social and solidarity economy (SSE) is emerging as a promising sector for the creation of inclusive jobs in urban areas. Driss Khrouz and Aicha Belarbi (2021, p. 278) examine the development potential of cooperatives and social enterprises in Moroccan cities. The authors emphasize the importance of appropriate regulatory frameworks and specific financing mechanisms to stimulate the growth of this sector, which is particularly effective in promoting the professional integration of vulnerable populations.

Promoting local entrepreneurship is emerging as a priority in employment support policies. Mohammed

Tozy and Beatrice Hibou (2020, p. 301) analyze initiatives to promote entrepreneurial culture and support business start-ups in the kingdom's main cities. Their research highlights the importance of incubators, coworking spaces and mentoring programs in stimulating the creation of innovative, job-creating businesses.

Public-private partnerships (PPPs) are emerging as a strategic tool for local job creation. Nadia Benabdejlil and Said Hanchane (2022, p. 245) examine the impact of major PPP infrastructure and urban development projects on local employment. The authors stress the importance of incorporating social clauses and local employment objectives into these contracts to maximize the economic benefits for urban communities.

Vocational training tailored to local needs appears to be an essential pillar of employment support policies. Abdellatif Miraoui and Khalid Soudi (2021, p. 189) analyze customized training initiatives set up in partnership with local businesses in several Moroccan cities. Their study highlights the positive impact of these programs on young people's employability and the match between skills supply and demand on the local labor market.

Policies to develop clusters and industrial ecosystems are emerging as major levers for the creation of skilled jobs. Hassan Zaoual and Fatima Zahra El Malki (2020, p. 234) examine clustering strategies implemented in growth sectors such as automotive, aeronautics and information technology. The authors highlight the importance of these approaches in creating local synergies, attracting investment and generating high value-added jobs.

The green economy and ecological transition are emerging as major sources of urban jobs. Fatima Arib and Mohammed Behnassi (2021, p. 312) analyze the potential for job creation in sectors such as renewable energies, energy efficiency and the circular economy in Moroccan cities. Their research highlights the importance of an incentive framework and strategic planning to stimulate the development of these green sectors that generate sustainable employment.

In conclusion, local policies to support job creation in Moroccan cities require a holistic and adaptive approach. As emphasized by Nadia El Ghazouani and Larabi Jaidi (2022, p. 335), these initiatives must be part of a global vision of urban economic development, combining the stimulation of supply, the development of skills and the improvement of the business environment. The authors stress the importance of participative governance and effective coordination between the various local players to maximize the impact of these policies. The challenge lies in the ability of Moroccan cities to create a dynamic and inclusive ecosystem, capable of generating diversified, high-quality employment opportunities for a growing urban population. The success of these local employment policies will be crucial in ensuring social cohesion, stimulating economic development and strengthening the resilience of Moroccan cities to the challenges of globalization and technological change.

### 3.3.2 Promoting entrepreneurship and supporting start-ups

Promoting entrepreneurship and supporting start-ups are key strategies for stimulating innovation, job creation and economic dynamism in Moroccan cities. These initiatives aim to create a favorable ecosystem for the emergence and development of new, innovative companies, capable of contributing to the diversification and modernization of the urban economic fabric.

According to an in-depth analysis by Rachid Guerraoui and Ghita Lahlou (2019, p. 178), Morocco has seen a significant evolution in its entrepreneurial ecosystem over the past decade, particularly in the major metropolises. The authors stress the importance of an integrated approach, combining incentive-based public policies, private sector involvement and the involvement of academic institutions to create an environment conducive to innovative entrepreneurship.

The development of support structures appears to be a key element in supporting the creation and growth of start-ups. Fatima Zahra El Malki and Hassan Zaoual (2020, p. 213) examine the impact of incubators and gas pedals on the survival and development of innovative start-ups in several Moroccan cities. Their study highlights the importance of personalized support, including mentoring, training and networking, in maximizing the chances of success for entrepreneurial projects.

Access to financing is a crucial issue for the development of start-ups. Nabil El Mabrouki and Karim Hasnaoui (2021, p. 245) analyze the various financing mechanisms available to Moroccan entrepreneurs, from business angels to venture capital funds. The authors stress the importance of developing a diversified financial ecosystem adapted to the different development phases of start-ups, while calling for greater involvement of institutional investors in financing innovation.

Promoting an entrepreneurial culture, particularly among young people, is a key priority. Mohammed Bensaid and Aïcha Belarbi (2020, p. 301) examine initiatives to integrate entrepreneurship into educational curricula and stimulate the entrepreneurial spirit among Moroccan students. Their research highlights the importance of such programs in changing mindsets and encouraging entrepreneurial risk-taking, particularly in a cultural context traditionally oriented towards salaried employment.

The development of sector-based ecosystems appears to be an important lever for stimulating innovative entrepreneurship. Abdellatif Miraoui and Khalid Soudi (2022, p. 189) analyze the clustering and smart specialization strategies implemented in several Moroccan cities to foster the emergence of start-ups in growth sectors such as information technology, biotechnology and renewable energies. The authors emphasize the importance of creating synergies between academic research, large companies and start-ups to stimulate innovation and value creation.

The internationalization of Moroccan start-ups is becoming a strategic issue for their growth and competitiveness. Nadia Benabdejlil and Said Hanchane (2021, p. 267) examine the international support programs set up to support the expansion of young Moroccan innovative companies. Their study highlights the importance of a comprehensive approach, including intercultural training, networking with foreign partners and support for adapting products to international markets.

Promoting female entrepreneurship is emerging as an important area for unlocking the entrepreneurial potential of Moroccan cities. Fatima Sadiqi and Moha Ennaji (2020, p. 234) analyze the specific initiatives put in place to support women entrepreneurs, particularly in terms of access to financing and mentoring. The authors underline the importance of these programs in promoting economic inclusion and empowerment of women in the urban entrepreneurial fabric.

In conclusion, promoting entrepreneurship and supporting start-ups in Moroccan cities requires an ecosystemic and collaborative approach. As Driss Ksikes and Kenza Sefrioui (2022, p. 335) emphasize, these initiatives must be part of a long-term vision of urban economic development, combining innovation, training and international openness. The authors stress the importance of creating dynamic "entrepreneurial communities", capable of generating positive emulation and fostering the sharing of experience between entrepreneurs. The challenge lies in the ability of Moroccan cities to create an environment conducive to the emergence and growth of innovative start-ups, capable of positioning themselves on national and international markets. The success of these policies to promote entrepreneurship will be crucial in stimulating innovation, creating skilled jobs and boosting the competitiveness of Morocco's urban economic ecosystems in the face of the challenges of the fourth industrial revolution.

### 3.3.3 Development of incubators and coworking spaces

The development of incubators and coworking spaces represents a strategic lever for stimulating entrepreneurship and innovation in Moroccan cities. These structures play a crucial role in the entrepreneurial ecosystem, providing an environment conducive to the emergence and growth of innovative young businesses.

Incubators, in particular, provide personalized support for project holders and early-stage start-ups. They generally offer a range of services including workspace, mentoring, training, access to professional networks and sometimes even financial support. As

Mohammed El Hassani (2022, p. 78) points out in his analysis of the Moroccan entrepreneurial landscape, "incubators play a catalytic role in reducing the failure rate of young businesses and accelerating their development through an integrated support ecosystem".

At the same time, coworking spaces have seen remarkable growth in Moroccan cities in recent years. These shared workplaces offer a flexible, cost-effective alternative to traditional offices, while fostering interaction and synergies between professionals from diverse backgrounds. In their study on the impact of coworking spaces on urban innovation in Morocco, Fatima Zahra Benali and Youssef Moussaoui (2023, p. 112) observe that "these spaces act as creative hubs, stimulating interdisciplinary collaboration and the emergence of innovative ideas within the local economic fabric".

The development of these structures is part of a broader strategy to create dynamic entrepreneurial ecosystems in Moroccan cities. Rachid Abdelkader (2021, p. 205) highlights the role of local authorities in this process: "Moroccan municipalities play a facilitating role by making premises available, simplifying administrative procedures and creating public-private partnerships to support the creation of incubators and coworking spaces".

However, challenges persist in the deployment of these structures. Nadia El Bouâzzaoui (2023, p. 167) points to "the need for better coordination between the various players in the entrepreneurial ecosystem to maximize the impact of incubators and coworking spaces on local economic development". In particular, it recommends a more targeted approach based on the sectoral and territorial specificities of each city.

What's more, the integration of new technologies into these spaces appears to be a crucial issue in boosting their attractiveness and efficiency. Karim Benjelloun (2022, p. 93) stresses the importance of "equipping incubators and coworking spaces with state-of-the-art digital infrastructures to meet the needs of technology start-ups and foster digital innovation".

In conclusion, the development of incubators and coworking spaces is proving to be a strategic way of strengthening the economic resilience of Moroccan cities. These structures help to create an environment conducive to entrepreneurship, innovation and collaboration, which are essential for boosting the local economic fabric and attracting talent. Their deployment, however, requires a concerted approach tailored to local realities to maximize their impact on sustainable urban development.

### 3.3.4 Support for the social economy and urban cooperatives

Supporting the social and solidarity economy (SSE) and urban cooperatives is a key way of strengthening the economic resilience of Moroccan cities. This sector, characterized by alternative modes of production and consumption, plays a crucial role in job creation, social inclusion and sustainable local development.

The social and solidarity economy in Morocco has experienced significant growth in recent years, driven by a growing awareness of social and environmental issues. As Amina Lamrani (2022, p. 45) points out in her analysis of the SSE landscape in Morocco, "the sector represents considerable potential for meeting the socio-economic challenges of cities, particularly in terms of creating inclusive jobs and local services".

Urban cooperatives, in particular, are emerging as key players in this movement. They offer an economic model based on resource pooling and democratic governance, particularly suited to urban contexts. Hassan El Ouazzani (2023, p. 132) observes that "urban cooperatives contribute to the revitalization of neighborhoods by creating bonds of solidarity and offering services adapted to local needs, from food production to waste management".

The support of local authorities is crucial to the development of SSE and urban cooperatives. Rachida Benali (2021, p. 78) highlights several levers for action: "Municipalities can play a facilitating role by making premises available, integrating social clauses into public procurement contracts, and creating specific support schemes for SSE structures".

However, challenges persist in the development of this sector. Karim Mounir (2022, p. 210) points to "the need to strengthen the management and marketing skills of SSE players to ensure their economic sustainability". He advocates the introduction of appropriate training programs and the development of partnerships with the traditional private sector.

Access to financing also remains a major challenge for SSE structures and urban cooperatives. Fatima Zohra El Malki (2023, p. 95) stresses the importance of "developing innovative financial instruments, such as social impact investment funds or participatory financing platforms, to meet the specific needs of these organizations".

The integration of new technologies into SSE models also appears to be a promising area for development. Youssef Lahlou (2022, p. 167) highlights the potential of "collaborative digital platforms to strengthen urban solidarity networks and optimize the distribution of goods and services produced by cooperatives".

In conclusion, support for the social and solidarity economy and urban cooperatives is a relevant strategy for strengthening the economic resilience of Moroccan cities. This sector offers innovative solutions to urban social and environmental challenges, while promoting a more inclusive and sustainable development model. However, its deployment requires a concerted approach between local authorities, the private sector and civil society, as well as ongoing adaptation to technological and societal developments.

### 3.3.5 Economic inclusion of vulnerable populations and young people

The economic inclusion of vulnerable populations and young people is a major challenge for the resilience and sustainable development of Moroccan cities. This issue, at the heart of the country's socio-economic concerns, requires a multidimensional and concerted approach to overcome the structural obstacles to integrating these groups into the urban economic fabric.

Vulnerable populations, including people in precarious situations, marginalized women and people with disabilities, face significant barriers in their access to employment and economic opportunities. As Nadia El Fassi (2022, p. 87) points out in her study of urban social exclusion in Morocco, "the mechanisms of economic exclusion are mutually reinforcing, creating vicious circles of poverty and marginalization in disadvantaged urban areas". This observation highlights the need for targeted, systemic interventions to break these cycles.

At the same time, the question of the economic integration of young people is becoming increasingly acute in a context of high unemployment and a mismatch between training and the needs of the job market. According to Mohammed Bougroum (2023, p. 156), "the challenge of youth employment in urban areas calls for a thorough overhaul of education and vocational training systems, as well as active employment policies tailored to local realities".

To meet these challenges, we have identified several areas for action. Firstly, strengthening vocational training and qualification programs is a priority. Fatima Zahra Alaoui (2021, p. 213) recommends "setting up flexible, modular training programs, including internships and entrepreneurship modules, to better prepare young people and vulnerable populations for the demands of the job market".

Social entrepreneurship and the solidarity economy also offer promising prospects for economic inclusion. Rachid Benmokhtar (2022, p. 129) points out that "social entrepreneurship initiatives, particularly those led by young people, can play a catalytic role in creating inclusive jobs and solving local social problems". It recommends that specific support and funding mechanisms be set up to support these initiatives.

Access to financing remains a crucial issue for the economic integration of vulnerable populations and young people. Samira El Morabet (2023, p. 78) highlights the importance of "developing adapted financial products, such as microcredit or guarantee funds, to facilitate access to capital for entrepreneurs from disadvantaged backgrounds". It also highlights the potential role of financial technologies (fintech) in democratizing access to financial services.

The fight against discrimination and stereotyping in the job market is also a key area for action. Karim

Bensaid (2022, p. 195) insists on "the need to raise awareness among employers and put in place incentive mechanisms to encourage the hiring of people from marginalized groups".

Urban policies also play a crucial role in economic inclusion. Leila Bouasria (2023, p. 241) argues in favor of an "integrated approach to urban development, combining interventions on housing, mobility and public services, to create an environment conducive to the economic insertion of vulnerable populations".

Finally, exploiting the potential of digital technology appears to be a promising lever for economic inclusion. Youssef El Karimi (2022, p. 167) points out that "digital platforms and the collaborative economy offer new employment and entrepreneurial opportunities, particularly attractive to urban youth". However, he recommends that this digital transition be accompanied by appropriate social protection measures.

In conclusion, the economic inclusion of vulnerable populations and young people in Moroccan cities requires a holistic approach, combining interventions in education, training, entrepreneurship, access to financing and the fight against discrimination. This approach must be part of a long-term vision of sustainable urban development, taking into account local specificities and technological developments. The success of these inclusion policies is crucial not only for social cohesion, but also for the competitiveness and economic resilience of Moroccan cities in a changing global context.

## 3.4 Local finance, budget management and resource mobilization

### 3.4.1 Greater financial autonomy for local authorities

Strengthening the financial autonomy of local authorities is crucial to the development and economic resilience of Moroccan cities. This autonomy is essential to enable municipalities to respond effectively to the needs of their populations and implement development strategies tailored to their specific contexts.

As Mohammed El Bakkali (2022, p. 112) points out in his analysis of financial decentralization in Morocco, "the financial autonomy of local authorities is a fundamental pillar of territorial governance, enabling a more efficient allocation of resources and a better match between public policies and local realities". This observation highlights the importance of increasing cities' financial leeway to strengthen their capacity for action.

The Moroccan legal framework has undergone significant changes in recent years to promote this autonomy. Nadia Benali (2023, p. 78) notes that "the organic law on the regions and the texts implementing advanced regionalization have laid the foundations for greater financial autonomy for local authorities".

However, the effective implementation of these provisions remains a major challenge.

One of the priorities for strengthening the financial autonomy of Moroccan cities is to diversify and optimize local revenue sources. Rachid Zouaoui (2021, p. 195) calls for "a review of the local taxation system to broaden the tax base and improve tax collection". He also stresses the importance of developing effective equalization mechanisms to reduce inequalities between territories.

Enhancing the value of local authorities' land and property assets is another promising lever. Fatima Zahra El Malki (2022, p. 143) highlights "the potential of public-private partnerships and innovative financial arrangements to generate recurrent income from municipal assets". However, she recommends a cautious and transparent approach to the management of these partnerships to safeguard the public interest.

Access to financial markets is also a key factor in strengthening cities' financial autonomy. Karim Benjelloun (2023, p. 210) points out that "issuing municipal bonds and using other financial instruments can offer local authorities alternative sources of funding for their investment projects". However, he insists on the need to strengthen municipalities' financial management and credit rating capabilities.

Digitizing municipal services and optimizing financial management appear to be complementary levers. Samira Ouazzani (2022, p. 167) observes that "the adoption of digital solutions for tax collection and budget management can significantly improve the efficiency and transparency of local finances". She recommends sustained investment in training for municipal staff and in digital infrastructures.

It is also crucial to strengthen the planning and financial management capabilities of local authorities. Hassan El Mouden (2021, p. 89) stresses "the importance of developing financial forecasting and multi-year investment management tools to improve the sustainability of local finances". He recommends the introduction of ongoing training programs for elected representatives and municipal managers.

Finally, the question of coordination between different levels of government remains central. Leila Bouasria (2023, p. 231) points out that "strengthening the financial autonomy of local authorities requires a clarification of powers and responsibilities between the central state and local authorities". She calls for greater dialogue and contractualization of financial relations between the various administrative levels.

In conclusion, strengthening the financial autonomy of Moroccan local authorities appears to be a complex and multidimensional process. It involves not only legal and fiscal reforms, but also the strengthening of local capacities, the modernization of management tools, and the evolution of relations between different levels of government. This increased autonomy is essential to enable Moroccan cities to meet the challenges of sustainable development and economic inclusion, while adapting to the specific characteristics of their territories. The success of this transformation requires a gradual and concerted approach, involving all stakeholders in a shared vision of territorial development.

## 3.4.2 Optimizing local taxation and broadening the tax base

Optimizing local taxation and broadening the tax base are crucial to strengthening the financial autonomy and capacity for action of Moroccan local authorities. These objectives are part of a broader approach to modernizing territorial governance and local economic development.

As Abdelkader Kaioua (2022, p. 87) points out in his analysis of local finances in Morocco, "local taxation represents an essential lever for territorial development, enabling local authorities to finance their projects and provide quality public services". This observation highlights the strategic importance of efficient and equitable local taxation.

One of the main challenges is to broaden the tax base. Fatima Zahra Benali (2023, p. 132) notes that "many economic activities still escape local taxation, particularly in the informal sector". She advocates "an approach combining incentives for formalization and strengthening local authorities' tax control capacities".

The modernization of local tax collection and management tools appears to be a priority. Rachid El

Houdaigui (2021, p. 205) highlights "the potential of digital technologies to improve taxpayer identification, facilitate collection and reduce tax evasion". In particular, he recommends "the implementation of geographic information systems to optimize the management of property tax and council tax".

Optimizing tax rates is another important lever. Nadia El Fassi (2022, p. 168) stresses the need for "a balance between the economic attractiveness of the territory and the financing needs of local authorities". She proposes "a differentiated approach according to business sectors and geographical areas, to take account of local economic disparities".

The question of fiscal equalization between territories remains central. Karim Bensaid (2023, p. 91) stresses "the importance of effective redistribution mechanisms to reduce territorial inequalities and ensure balanced development on a national scale". He advocates "a revision of the criteria for distributing State subsidies to take better account of the specific burdens of local authorities".

Improving transparency and communication around local taxation also appears to be a major challenge. Samira Ouazzani (2021, p. 173) observes that "citizens' understanding and acceptance of local taxes are essential to improving fiscal civic-mindedness". She recommends "setting up awareness-raising campaigns and citizen consultation mechanisms on the use of tax revenues".

Training and capacity-building for local tax authorities are priorities. Hassan El Mouden (2022, p. 218) stresses "the need to invest in the ongoing training of tax agents and in high-performance management tools to improve collection efficiency". He proposes "the creation of training centers specialized in local taxation at regional level".

Innovation in green taxation and tax incentives is a promising avenue. Leila Bouasria (2023, p. 145) highlights "the potential of environmental taxes and incentive mechanisms to steer economic behavior towards more sustainable development". In particular, she cites "the successful experience of certain cities in introducing taxes on plastic packaging or tax incentives for energy-efficient building renovation".

Finally, coordination between local and national taxation is a crucial issue. Youssef Lamrani (2022, p. 197) insists on "the need for better articulation between the different levels of taxation to avoid double taxation and optimize the distribution of resources". He calls for "a comprehensive reform of the Moroccan tax system, fully integrating the territorial dimension".

In conclusion, optimizing local taxation and broadening the tax base in Morocco requires a multidimensional approach, combining legal reform, technological innovation, capacity building and improved governance. These efforts are essential to provide local authorities with the financial resources they need to develop and improve public services. The success of this fiscal transformation requires close collaboration between the various levels of government, the private sector and civil society, with a view to sustainable and inclusive territorial development.

### 3.4.3 Efficient expense management and rationalization of investments

Efficient management of expenditure and rationalization of investments are major challenges for Moroccan local authorities, in a context of limited resources and growing demand for public services and urban infrastructure. This issue is at the heart of efforts to modernize local governance and optimize the use of public funds.

As Mohammed El Khadiri (2022, p. 112) points out in his analysis of the financial management of Moroccan cities, "efficiency in the allocation and use of public resources has become an inescapable imperative to ensure the sustainability of local finances and meet citizens' expectations". This observation highlights the importance of a strategic and rigorous approach to managing municipal expenditure.

One of the priorities is to improve budget planning. Nadia Benali (2023, p. 78) advocates "the widespread adoption of multi-year, performance-based budgeting to better align spending with long-term development objectives". She stresses the importance of integrating clear, measurable performance indicators into the budgeting process.

The rationalization of operating expenses appears to be an important lever for freeing up financial leeway. Rachid Zouaoui (2021, p. 195) highlights "the potential of organizational audits and the pooling of services between local authorities to optimize operating costs". In particular, he recommends "setting up joint purchasing groups to benefit from economies of scale".

Optimizing the management of local authorities' real estate assets is another area for action. Fatima Zahra El Malki (2022, p. 143) points out that "active management of municipal real estate assets can generate substantial savings and additional revenue". She suggests "drawing up real estate master plans to rationalize the use of public buildings and enhance the value of under-utilized assets".

Improving the energy efficiency of urban infrastructure represents a promising avenue for reducing expenditure in the long term. Karim Benjelloun (2023, p. 210) observes that "investments in the energy renovation of public buildings and the modernization of urban lighting can generate significant savings on operating costs". He calls for "the systematic adoption of energy efficiency criteria in public procurement contracts".

The prioritization and rigorous evaluation of investment projects are crucial issues. Samira Ouazzani (2022, p. 167) insists on "the need to strengthen cost-benefit analysis and ex-ante project evaluation capacities within local administrations". She recommends "setting up multidisciplinary investment committees to ensure objective project selection".

Improving transparency and control in public expenditure management appears to be a key priority. Hassan El Mouden (2021, p. 89) stresses that "the strengthening of internal and external control mechanisms, as well as the regular publication of reports on budget execution, are essential to improve spending efficiency". He advocates "the adoption of digital financial management platforms to facilitate real-time monitoring of expenditure".

Involving citizens in the budgeting process represents an innovative approach to improving spending efficiency. Leila Bouasria (2023, p. 231) highlights "the potential of participatory budgets and citizen consultations to better align investments with the real needs of the population". She cites the successful experiences of some Moroccan cities in setting up participatory budget democracy mechanisms.

Finally, the development of financial management skills within local administrations is crucial. Youssef El Karimi (2022, p. 167) stresses "the importance of ongoing training programs for elected representatives and municipal managers in budget management and financial analysis". He proposes "the creation of networks for the exchange of best practices between local authorities to encourage mutual learning".

In conclusion, effective management of

expenditure and rationalization of investment in Morocco's local authorities require a comprehensive approach, combining strategic planning, operational optimization, tighter controls and citizen involvement. These efforts are essential to improve the quality of public services and support local development, in a context of budgetary constraints. The success of this transformation requires a strong commitment from elected representatives, increased professionalization of local administrations, and a culture of performance and transparency in the management of public funds.

### 3.4.4 Public-private partnerships to finance urban projects

Public-private partnerships (PPPs) for financing urban projects represent a strategic development lever for Moroccan cities, offering opportunities to mobilize resources and expertise to meet infrastructure and public service challenges. This approach is part of a drive to modernize urban governance and find innovative solutions to local authorities' budgetary constraints.

As Mohammed El Hassani (2022, p. 118) points out in his analysis of PPPs in Morocco, "public-private partnerships offer considerable potential for accelerating the implementation of structuring urban projects, while optimizing the allocation of risks between the public and private sectors". This observation highlights the growing interest in these arrangements in the Moroccan context.

The legal framework for PPPs in Morocco has evolved significantly in recent years. Nadia Fassi Fihri (2023, p. 85) notes that "Law 86-12 on public-private partnership contracts laid the foundations for a regulatory framework conducive to the development of PPPs, by clarifying procedures and guarantees for investors". However, it stresses the need to adapt this framework to the specific characteristics of local authorities.

One of the main advantages of PPPs lies in their ability to mobilize private financing for projects of public interest. Rachid Zouaoui (2021, p. 203) observes that "PPPs enable cities to make large-scale investments without immediately increasing their debt, while benefiting from the technical and managerial expertise of the private sector". However, he warns against the risk of long-term debt overload, and recommends rigorous analysis of the financial sustainability of each project.

The diversity of PPP models offers considerable flexibility to meet the specific needs of urban projects. Fatima Zahra Benali (2022, p. 157) points out that "concession contracts, delegated management contracts or mixed economy companies make it possible to adapt the sharing of responsibilities and risks according to the nature of the project and the capacities of the partners". It recommends a tailor-made approach for each project, based on an in-depth analysis of local needs and constraints.

Improving the governance of PPPs is crucial to their success. Karim Benjelloun (2023, p. 219) stresses "the importance of strengthening local authorities' capacity to structure, negotiate and monitor PPP contracts". He advocates "the creation of specialized units within municipal administrations and the use of external expertise for complex projects".

Transparency and social acceptability of PPPs are major challenges. Samira Ouazzani (2021, p. 175) highlights "the need to involve citizens and local stakeholders in the process of designing and implementing PPPs, to ensure that they meet the real needs of the population". It recommends "organizing public consultations and setting up participatory project monitoring mechanisms".

Innovation in PPP financing models is a promising avenue. Hassan El Mouden (2022, p. 231) highlights the potential of "social or environmental impact bonds to finance urban projects with high societal added value". He cites the example of certain Moroccan cities that have experimented with these instruments for social housing or waste management projects.

The link between PPPs and sustainable urban development strategies is a central issue. Leila Bouasria (2023, p. 188) stresses "the importance of integrating sustainability and resilience criteria into the design and evaluation of PPP projects". She advocates "the systematic adoption of environmental and social impact assessments for all major projects".

Finally, the development of a local ecosystem conducive to PPPs appears to be a prerequisite for their success. Youssef Lamrani (2022, p. 204) highlights "the role of incubators, specialized investment funds and matchmaking platforms in stimulating the emergence of innovative PPP projects at local level". He recommends "the creation of regional centers of expertise to support local authorities in setting up and monitoring PPPs".

In conclusion, public-private partnerships offer significant opportunities for financing and implementing urban projects in Morocco, while presenting challenges in terms of governance, transparency and balance of interests. Their development requires a cautious and thoughtful approach, combining an appropriate regulatory framework, local capacity building, citizen involvement and a long-term vision of urban development. The success of PPPs in Moroccan cities will depend on the ability to reconcile economic efficiency, social responsibility and environmental sustainability in project design and implementation.

## 3.4.5 Access to innovative financing and green finance

Access to innovative financing and green finance represents a crucial challenge for Moroccan cities, in a context where the need for sustainable urban investment is considerable and traditional resources are often limited. This approach is part of a global drive towards ecological transition and the search for financial solutions tailored to the challenges of sustainable urban development.

As Nadia El Fassi (2022, p. 87) points out in her analysis of new forms of urban financing, "green finance mechanisms and innovative financial instruments offer Moroccan cities unprecedented opportunities to finance their ecological transition and climate change adaptation projects". This observation highlights the transformative potential of these approaches for sustainable urban development.

One of our priorities is the development of green bonds and social impact bonds. Mohammed Rachid (2023, p. 132) notes that "green bonds enable local authorities to raise funds dedicated specifically to environmental projects, while often benefiting from advantageous financial conditions". He cites the pioneering example of the city of Casablanca, which has successfully issued green bonds to finance energy efficiency and sustainable mobility projects.

Crowdfunding is emerging as a promising source of funding for local-scale urban projects. Fatima Zahra Benali (2021, p. 195) observes that "crowdfunding platforms make it possible to mobilize citizens' savings for projects of general interest, thereby strengthening community involvement and transparency in the management of urban projects". However, it recommends a suitable regulatory framework to secure these practices.

Investment funds specializing in sustainable urban development represent another interesting avenue. Karim Bensaid (2022, p. 218) highlights "the potential role of impact funds and green funds in channelling private investment into sustainable urban projects". He stresses the importance of developing blending finance mechanisms, combining public and private funds to reduce risk and attract investors.

Access to international climate financing is a major challenge for Moroccan cities. Samira Ouazzani (2023, p. 156) insists on "the need to build the capacity of local authorities to access climate funds such as the Green Climate Fund or the Adaptation Fund". She advocates "the creation of specialized units within municipal administrations to set up and monitor projects eligible for climate financing".

Payment for ecosystem services mechanisms are emerging as an innovative approach. Hassan El Mouden (2022, p. 173) points out that "these mechanisms make it possible to value and finance the preservation of urban and peri-urban ecosystems, which are essential for the resilience of cities". He cites the pilot experiences of some Moroccan cities in setting up remuneration systems for watershed protection.

Innovation in pricing models for urban services appears to be a lever for generating sustainable resources. Leila Bouasria (2021, p. 209) highlights "the potential of dynamic pricing systems and ecological equalization mechanisms to finance the transition to more sustainable urban infrastructures". However, she recommends a cautious approach to ensuring social equity in access to essential services.

Developing partnerships with the banking sector to promote green financial products is a key strategy. Youssef Lamrani (2023, p. 241) observes that "collaboration between local authorities and banks can foster the emergence of financial products tailored to sustainable urban projects, such as green loans or impact credit lines". He stresses the importance of an incentive framework to encourage these financial innovations.

Finally, the integration of ESG (Environmental, Social and Governance) criteria into cities' financial strategies appears to be an emerging trend. Rachid Zouaoui (2022, p. 185) stresses "the importance of local authorities adopting extra-financial reporting practices to improve their attractiveness to responsible investors". He recommends training elected representatives and municipal managers in the challenges of sustainable finance.

In conclusion, access to innovative financing and green finance opens up promising prospects for Moroccan cities, enabling them to reconcile urban development, ecological transition and social inclusion. However, these approaches require a strengthening of local capacities, an adaptation of the regulatory framework and a long-term strategic vision. The success of this financial transition requires close collaboration between local authorities, the private sector, civil society and international financial institutions, with a view to sustainable and resilient urban development.

## Conclusion:

The economic resilience of Moroccan cities in a globalized context is a multidimensional and complex issue, requiring a holistic and adaptive approach. In-depth analysis of the various aspects discussed in this chapter highlights the interconnection between the economic, social, environmental and governance dynamics shaping urban development in Morocco.

Economic diversification and the development of new activities are emerging as key strategic axes for strengthening the resilience of urban economies. The ability of Moroccan cities to integrate themselves into global value chains, while making the most of their local assets, is a key factor in their long-term competitiveness. This observation highlights the importance of a balanced approach, combining international openness and territorial roots in local economic development strategies.

Investment attractiveness and territorial competitiveness are essential levers for boosting urban economies. However, attractiveness must not be achieved at the expense of the sustainability and inclusiveness of urban development. This reflection underlines the need for an integrated approach, reconciling economic imperatives, social cohesion and environmental preservation in development and territorial promotion policies.

Employment, entrepreneurship and the social and solidarity economy appear to be crucial vectors of economic inclusion and social resilience. The catalytic role of local entrepreneurial ecosystems in sustainable job creation and social innovation is particularly noteworthy. This perspective highlights the importance of public policies favoring the emergence of a diversified economic fabric rooted in local realities.

The question of local finances and resource mobilization is becoming a cross-cutting issue, conditioning cities' ability to implement their development strategies. Optimizing local taxation, rationalizing expenditure and accessing innovative sources of financing are major challenges. The financial autonomy of local authorities is inextricably linked to their ability to drive sustainable, resilient urban development. This observation highlights the need for in-depth reform of local financial governance.

A cross-cutting analysis of these different dimensions reveals the emergence of new paradigms in the design and implementation of urban policies in Morocco. The transition to more sustainable, inclusive and resilient development models requires a rethinking of traditional approaches, integrating the principles of circular economy, social innovation and participatory governance. The economic resilience of Moroccan cities depends on their ability to articulate global dynamics and local specificities, in a perspective of integrated territorial development.

In conclusion, the economic resilience of Moroccan cities in a globalized context appears to be a dynamic and multifaceted process, requiring ongoing adaptation of public policies and governance practices. The challenges identified call for a systemic approach, combining economic innovation, social inclusion, environmental sustainability and institutional capacity-building. The success of this urban transformation will depend on the ability of local players to mobilize the resources, skills and partnerships needed to meet contemporary and future challenges. This evolution towards more resilient and sustainable cities is part of a long-term dynamic, requiring ongoing commitment and a shared vision between the various stakeholders involved in urban development in Morocco.

# *Chapter 4: Social resilience and the fight against urban inequalities*

The fourth chapter of this study addresses the crucial issue of social resilience and the fight against urban inequalities, a fundamental challenge for the sustainable development of contemporary cities. In a global context marked by growing and often poorly controlled urbanization, socio-economic disparities within urban areas are tending to increase, threatening social cohesion and territorial stability. This phenomenon, observed in both developed and developing countries, calls for in-depth reflection on how to strengthen the capacity of cities to absorb shocks, adapt to change and maintain an inclusive, resilient social fabric.

The analysis of urban social resilience requires a multidimensional approach, integrating aspects as varied as access to basic services, participative governance, urban security and the enhancement of cultural heritage. This complexity reflects the systemic nature of the challenges facing contemporary cities, where social, economic, environmental and cultural issues are intimately intertwined. The fight against urban inequalities cannot therefore be limited to isolated sectoral interventions, but must form part of a global, integrated strategy for sustainable urban development.

This chapter explores the various facets of urban

social resilience in four main areas. Firstly, improving access to basic services and reducing urban poverty are essential to ensuring dignified living conditions for all inhabitants. Secondly, strengthening social cohesion and citizen participation is a major lever for building more resilient and inclusive urban communities. Thirdly, urban security and conflict prevention are seen as sine qua non conditions for harmonious social development. Finally, the enhancement of cultural heritage and local identity is examined as a potential factor in strengthening social ties and territorial attractiveness.

The aim of this chapter is to offer a critical analysis of the strategies and tools that can be mobilized to promote social resilience and combat urban inequalities. Drawing on concrete examples and case studies, it aims to identify innovative courses of action that can be adapted to different urban contexts. This reflection is part of the broader framework of the United Nations' Sustainable Development Goals (SDGs), in particular SDG 11, which aims to make cities inclusive, safe, resilient and sustainable places.

## 4.1 Access to basic services and urban poverty reduction

### 4.1.1 Improving access to decent, affordable housing

Improving access to decent, affordable housing is a key factor in combating urban inequalities and strengthening social resilience. This issue, which lies at the crossroads of urban, social and economic policies, is of crucial importance in a context of population growth and accelerated urbanization. As Jean-Claude Driant (2015, p. 23) points out in his book "Les politiques du logement en France", "housing is an essential asset, at once a support for family intimacy and a vehicle for social and professional integration". This dual private and social dimension of housing makes it a fundamental lever for inclusion and urban cohesion.

The question of accessibility to decent, affordable housing is particularly acute in major metropolises, where land and real estate pressure tends to exclude the most vulnerable populations from central, well-served areas. This phenomenon, described as "gentrification" by Anne Clerval (2013, p. 87) in her study "Paris sans le peuple: La gentrification de la capitale", contributes to reinforcing socio-spatial disparities and weakening the urban social fabric. Faced with these dynamics, public policies must strive to maintain a diversified and accessible housing supply throughout the urban territory, in order to preserve social mix and combat residential segregation phenomena.

Improving access to housing requires a combination of supply-side and demand-side measures. On the supply side, the production of social and intermediate housing plays a crucial role. As Pierre Merlin (2018, p. 156) explains in "L'urbanisme", "social housing policy aims to offer quality housing to households that would not be able to find decent accommodation on the open market". This approach requires a strong commitment from the public authorities, both in terms of financing and property market regulation. On the demand side, personalized housing assistance and social support schemes are in place to help low-income households access and maintain housing that meets their needs.

Urban renewal is also an important lever for improving the quality and accessibility of existing housing stock. Renaud Epstein (2013, p. 201), in his book "La rénovation urbaine : Démolition-reconstruction de l'État", analyzes the contrasting effects of urban renewal programs in France, highlighting both their potential to physically transform run-down neighborhoods and the risks of displacing the most fragile populations. These interventions must therefore be carried out with particular attention to their social impact, taking care to preserve existing solidarity networks and involve residents in the process of transforming their living environment.

Innovation in housing forms and production methods appears to be a promising avenue for meeting the challenges of accessibility and sustainability. Marie-Hélène Bacqué and Stéphanie Vermeersch (2019, p. 112), in their study "L'habitat participatif: De l'initiative habitante à l'action publique", highlight the potential of participatory housing approaches to produce affordable housing while strengthening social ties and citizen involvement. These alternative approaches, which also include cooperative housing and assisted self-building, open up new prospects for reconciling affordability, architectural quality and resident participation.

In conclusion, improving access to decent and affordable housing requires a comprehensive and integrated approach, articulating interventions on the built environment, land policies, financial arrangements and social support. As summarized by Yankel Fijalkow (2017, p. 78) in "Sociologie du logement", "the question of housing cannot be dissociated from broader issues of spatial justice and the right to the city". This perspective invites us to think about housing accessibility not only in quantitative terms, but also in qualitative terms, taking into account the location, environmental quality and urban integration of the housing produced.

## 4.1.2 Strengthening drinking water and sanitation infrastructures

Strengthening drinking water and sanitation infrastructures is a fundamental pillar of urban resilience and the fight against inequalities. This theme, at the crossroads of health, environmental and social issues, is of paramount importance in the sustainable development of cities. As Bernard Barraqué (2016, p. 45) points out in his book "L'eau des villes: aux sources des empires municipaux", "access to drinking water and sanitation is an essential marker of urban quality of life and a determining factor in public health".

The issue of access to drinking water and

sanitation is particularly acute in rapidly expanding urban areas, especially in developing countries. Julien Custot and Agathe Euzen (2017, p. 78), in their study "L'accès à l'eau en ville : enjeux techniques et sociaux du Sud au Nord", highlight the considerable challenges faced by local authorities in extending and maintaining water and sanitation networks in the face of often uncontrolled urban growth. The authors stress the importance of an integrated approach, taking into account not only the technical aspects, but also the social and cultural dimensions of access to water.

Sustainable management of water resources in urban environments requires a systemic approach, integrating the various stages of the water cycle. As Jean-Claude Deutsch (2015, p. 132) explains in "L'eau dans la ville : Une amie qui nous fait la guerre", "the city must be thought of as a complex hydraulic ecosystem, where stormwater management, drinking water supply and wastewater treatment are intimately linked". This holistic vision means rethinking urban planning to encourage the natural infiltration of rainwater, reduce the risk of flooding and optimize the use of water resources.

Improving access to drinking water and sanitation also requires technological and organizational innovations. Laetitia Guérin- Schneider and Michel Nakhla (2018, p. 201), in their book "La gouvernance des services publics d'eau et d'assainissement", analyze the different models for managing and regulating water services, emphasizing the importance of transparent, participatory governance to guarantee the efficiency and equity of these essential services. The authors highlight the potential of digital technologies to optimize network management and improve relations with users.

The question of equity in access to water and sanitation lies at the heart of environmental justice issues in urban areas. As Catherine Baron (2014, p. 89) points out in her study "Water and development: a critical perspective", "access to drinking water and sanitation often reflects and reinforces existing socio-spatial inequalities within cities". This observation calls for particular attention to be paid to precarious and informal neighborhoods, where infrastructure deficits are generally most pronounced. Innovative approaches, such as decentralized sanitation systems or alternative potabilization technologies, can offer solutions adapted to these specific urban contexts.

The environmental dimension of strengthening water and sanitation infrastructures must not be overlooked. Ghislain de Marsily (2019, p. 156), in his book "L'eau, un trésor en partage", stresses the need to adopt an ecosystem approach to urban water management, preserving aquatic environments and promoting the reuse of treated wastewater. This approach is part of a circular economy aimed at reducing the ecological footprint of cities and strengthening their resilience in the face of climate change.

In conclusion, strengthening drinking water and wastewater infrastructures is an essential lever for improving the quality of urban life and reducing inequalities. As Bruno Tassin (2020, p. 234) sums up in "L'eau dans la ville de demain", "the challenge is not only technical, but also social and political: it's a question of guaranteeing a fundamental right while preserving a precious resource". This multi-dimensional approach invites us to rethink the place of water in the city, not just as an urban service, but as a structuring element of sustainable, inclusive urban development.

### 4.1.3 Developing local health services

The notion of local health services is part of a logic of territorialization of health policies. According to Anne-Cécile Hoyez (2019, p. 45), "proximity in healthcare is not simply a question of geographical distance, but also encompasses social, cultural and organizational dimensions". This multidimensional approach implies a reflection on the accessibility of care, in both physical and socio-cultural terms.

One of the main aims of local healthcare services is to reduce territorial inequalities in health. As Emmanuelle Faure et al (2017, p. 112) point out, "disparities in access to healthcare between neighborhoods in the same city can be as marked as those observed between regions". The development of community health centers, multi-disciplinary medical centers and outpatient clinics in priority neighborhoods aims to bridge these gaps.

Setting up local health services requires an intersectoral approach. Pierre Lombrail and Thierry Lang (2020, p. 78) assert that "the effectiveness of these schemes depends on their ability to integrate the social determinants of health and to collaborate with other local players, such as social, educational or urban planning services". This holistic vision of urban health implies close coordination between different professionals and institutions.

Involving local residents in the design and management of community health services is also crucial. Zoé Vaillant (2018, p. 203) points out that "citizen participation makes it possible not only to adapt services to the real needs of the population, but also to strengthen community empowerment and health literacy". Initiatives such as city health workshops or local mental health councils illustrate this participatory approach.

Digital technology is playing a growing role in the development of local healthcare services. According to Gérard Reach (2021, p. 156), "telemedicine and mobile health applications offer new possibilities for bringing care closer to patients, particularly in underserved urban areas". However, he warns against the risk of digital exclusion and stresses the importance of maintaining human interaction in the care pathway.

Training and retaining healthcare professionals in

priority neighborhoods remains a major challenge. As noted by Yann Bourgueil et al (2016, p. 89), "the attractiveness of sensitive urban areas for doctors and other healthcare professionals depends not only on financial incentives, but also on the quality of the working environment and professional development prospects". Initiatives such as public service commitment contracts and multi-professional health centers aim to meet this challenge.

Finally, the ongoing evaluation and adaptation of local health services is essential to guarantee their relevance and effectiveness. Didier Fassin (2018, p. 231) points out that "the impact of these schemes must be measured not only in terms of health indicators, but also in terms of their ability to reduce social inequalities in health and improve residents' quality of life".

In conclusion, the development of local health services represents an important lever for improving social resilience and combating urban inequalities. This approach requires an integrated vision of urban health, intersectoral collaboration and the active involvement of local communities.

## 4.1.4 Improving access to education and combating early school leaving

Access to quality education is a fundamental right and a powerful vector of social mobility. However, as Marie Duru-Bellat (2017, p. 23) points out, "educational inequalities in urban areas are not simply a reflection of social inequalities, but actively contribute to reproducing and sometimes amplifying them". This observation highlights the crucial importance of education policies in combating urban inequalities.

Dropping out of school, defined as the premature interruption of an educational pathway, is a complex phenomenon with multiple causes. Pierre-Yves Bernard (2019, p. 87) identifies several risk factors: "learning difficulties, lack of motivation, family problems, economic insecurity, and the mismatch between the expectations of the educational institution and the socio-cultural realities of the pupils". This multiplicity of causes calls for diversified responses adapted to local contexts.

To improve access to education, numerous initiatives have been set up in priority urban areas. Priority Education Networks (REP and REP+) are an emblematic example. As Catherine Moisan (2018, p. 156) explains, "these schemes aim to concentrate human and material resources in schools catering for the most disadvantaged pupils, in order to reduce the achievement gap with the rest of the territory". However, the effectiveness of these measures is the subject of debate, with some critics pointing to their potential stigmatizing effect.

The fight against dropping out of school also involves the development of alternative and inclusive pedagogies. According to Philippe Meirieu (2020, p. 112), "it is crucial to diversify pedagogical approaches to adapt to different learning profiles and restore meaning to school knowledge". Experiments such as cooperative classes, project-based learning and Freinet pedagogy are showing promising results in engaging students at risk of dropping out.

Involving families in their children's educational journey is another important lever. Françoise Lorcerie (2016, p. 78) points out that "coeducation, i.e. close collaboration between school and parents, is particularly beneficial for pupils from disadvantaged backgrounds". Initiatives such as "parents' cafés" or school-family mediators aim to strengthen this crucial link.

Digital technology offers new opportunities for access to education and dropout prevention. Bruno Devauchelle (2019, p. 203) asserts that "digital tools can promote the personalization of learning paths and facilitate the monitoring of students in difficulty". However, he warns against the risk of widening the digital divide, and stresses the importance of supporting students and teachers in the use of these technologies.

Teacher training and support play a crucial role in the success of these policies. As Patrick Rayou (2017, p. 167) notes, "teaching in difficult neighborhoods requires specific skills in terms of classroom management, pedagogical differentiation and understanding socio-cultural issues". The development of in-service training and communities of practice between teachers can help reinforce these skills.

Finally, the territorial and partnership approach is essential to effectively combat school dropout. According to Choukri Ben Ayed (2018, p. 245), "dropout prevention cannot be limited to the school environment, but must involve all local players: social services, associations, local businesses". The Cités éducatives, launched in 2019, illustrate this desire to mobilize all local levers in the service of educational success.

In conclusion, improving access to education and combating school dropout in urban areas requires a comprehensive approach, mobilizing pedagogical, social and territorial resources. These efforts are essential if we are to break the cycles of inequality reproduction and foster greater social cohesion in cities.

### 4.1.5 Social support programs and safety nets for the most vulnerable

The notion of social safety nets encompasses a set of measures designed to guarantee a minimum standard of living for vulnerable populations. According to Robert Castel (2016, p. 34), "these schemes are part of a logic of collective solidarity and aim to prevent the social disaffiliation of the most fragile individuals". This approach recognizes that social vulnerability is often multidimensional, requiring coordinated and integrated responses.

In France, the social protection system plays a central role in providing these safety nets. As Bruno

Palier (2018, p. 112) points out, "the French social model, despite its limitations, has demonstrated its ability to cushion economic shocks and significantly reduce poverty". However, the author also notes that this system faces growing challenges, particularly in terms of financing and adapting to new forms of urban precariousness.

Social support programs in urban areas take many forms. One of the most emblematic is the Revenu de Solidarité Active (RSA). Nicolas Duvoux (2017, p. 78) analyzes the impact of this scheme: "The RSA has reduced the intensity of poverty for many urban households, but its effectiveness remains limited by problems of access to rights and non-recourse". This observation underlines the importance of simplifying administrative procedures and strengthening support for beneficiaries.

Housing is another major focus of policies to support the most vulnerable in urban areas. Julien Damon (2019, p. 156) notes that "access to decent, affordable housing is an essential pillar of social inclusion, particularly in large conurbations where real estate pressure is strong". Schemes such as social housing, housing subsidies or emergency accommodation play a crucial role in preventing homelessness and housing exclusion.

Health is also at the heart of social support programs. Didier Fassin (2020, p. 203) points out that "health inequalities in urban areas are closely linked to living conditions and social determinants". Schemes such as Couverture Maladie Universelle (CMU) and Aide Médicale d'Etat (AME) aim to guarantee access to healthcare for the most precarious populations, although their implementation sometimes remains problematic.

The professional integration of vulnerable people is another crucial issue. Emmanuèle Reynaud (2017, p. 132) analyzes the effectiveness of integration policies: "Support schemes to help people find work, such as integration projects or integration companies, play an important role in the reintegration of people who are far from employment, but their impact remains limited in the face of persistent structural unemployment". This observation underlines the need to combine integration policies with broader local economic development strategies.

Combating social isolation, which is particularly prevalent in urban areas, is also one of the objectives of support programs. Serge Paugam (2018, p. 245) highlights the importance of social ties: "Social mediation schemes, social centers or local solidarity networks play a crucial role in maintaining social ties and preventing the isolation of vulnerable people". These initiatives help to strengthen the urban social fabric and encourage community mutual aid.

The effectiveness of social support programs largely depends on their ability to adapt to local realities. Philippe Warin (2016, p. 189) stresses the importance of a territorialized approach: "The implementation of social policies must take into account the specific features of each urban territory, involving local players and promoting social innovation". This approach ensures a better match between the needs of the population and the responses provided.

In conclusion, social support programs and safety nets for the most vulnerable are an essential pillar of social resilience in urban areas. Their effectiveness relies on a comprehensive approach, integrating the economic, social and territorial dimensions of vulnerability. As Martin Hirsch (2020, p. 312) sums up, "the fight against poverty and exclusion in urban areas requires the mobilization of society as a whole, combining institutional solidarity and civic commitment".

## 4.2 Social cohesion, citizen participation and inclusive governance

### 4.2.1 Promoting inter-community dialogue and social diversity

The concept of intercommunity dialogue is part of a broader perspective on managing diversity in urban environments. As Michel Wieviorka (2017, p. 45) points out, "the challenge of contemporary societies is to reconcile the recognition of cultural differences with the maintenance of a common public space and shared values". This approach implies going beyond the logic of simple coexistence to foster genuine interactions and exchanges between communities.

Social diversity, for its part, is often presented as an ideal of French urban policies. However, its effective implementation raises many challenges. Marie-Hélène Bacqué and Eric Charmes (2016, p. 78) note that "social mix cannot be decreed; it is built over time and requires proactive policies that go beyond simple residential mix". This observation underlines the importance of a multidimensional approach that takes into account the economic, cultural and social aspects of diversity.

One of the major challenges of intercommunity dialogue is combating prejudice and discrimination. Patrick Simon (2018, p. 132) points out that "stereotypes and negative representations between communities are often the main obstacle to genuine dialogue". Initiatives such as intercultural weeks, citizen forums or collaborative art projects can help deconstruct these prejudices and foster greater mutual understanding.

The question of language plays a central role in intercommunity dialogue. Alexandra Filhon (2019, p. 203) points out that "mastery of the host country's language is a key factor in integration, but valuing languages of origin can also contribute to the recognition and self-esteem of immigrant communities". Schemes such as language courses for adults, multilingual libraries and intercultural writing workshops can foster this dual dynamic.

Promoting social diversity often involves housing

and urban planning policies. Renaud Epstein (2020, p. 167) analyzes the effects of the SRU law (Solidarité et Renouvellement Urbain): "While this law has made it possible to increase the proportion of social housing in many communes, its impact on effective diversity remains limited due to avoidance phenomena and micro-segregations at neighbourhood level". This observation underscores the need for housing policies to be accompanied by measures designed to promote living together on a daily basis.

Schools play a crucial role in promoting intercommunity dialogue and social mixing. Choukri Ben Ayed (2017, p. 245) asserts that "school can be a powerful vector for mixing and learning to live together, provided that appropriate pedagogies are put in place and school avoidance phenomena are combated". Initiatives such as school twinning, intercultural educational projects and peer mediation programs can all contribute to this goal.

Citizen participation is another important lever for promoting dialogue and diversity. Marion Carrel (2018, p. 112) points out that "participatory democracy mechanisms, when designed inclusively, can enable encounters and exchanges between groups that don't usually rub shoulders". Neighborhood councils, participatory budgets and citizens' juries are all potential spaces for intercommunity dialogue.

Finally, the role of associations and civil society initiatives is crucial in promoting dialogue and diversity. As Catherine Neveu (2016, p. 189) notes, "neighborhood associations, social centers or citizens' collectives often play a bridging role between communities, helping to forge links on a daily basis". Supporting these initiatives, both financially and logistically, is essential to fostering a sustainable dynamic of intercommunity dialogue.

In conclusion, promoting intercommunity dialogue and social mixing is a complex process that requires a comprehensive, long-term approach. As Dominique Schnapper (2020, p. 312) sums up, "the challenge is to build a society where diversity is recognized and valued, while maintaining a sense of common belonging and solidarity". This ambition calls for the mobilization of all urban players, from public authorities to citizens, associations and educational and cultural institutions.

### 4.2.2 Strengthening mechanisms for citizen participation in decision-making

Citizen participation is part of a broader evolution in modes of urban governance. As Loïc Blondiaux (2018, p. 23) points out, "the participatory imperative has become an unavoidable element of political discourse and public action, reflecting a growing demand for direct democracy on the part of citizens". This trend reflects a desire to go beyond the limits of traditional representative democracy.

There are many different forms of citizen participation, depending on the degree to which residents are involved. Marie-Hélène Bacqué and Mario Gauthier (2016, p. 45) propose a typology ranging from simple consultation to co-decision: "Between these two poles, we find a range of devices such as participatory budgets, citizen juries or consensus conferences, which offer varying degrees of power to citizens". This diversity makes it possible to adapt participatory mechanisms to local contexts and specific issues.

One of the major challenges of citizen participation is to ensure that the population is truly representative. Rémi Lefebvre (2019, p. 112) notes that "participatory mechanisms often tend to attract already politicized citizens and exclude the most marginalized groups". To overcome this bias, initiatives such as drawing lots or setting socio-demographic quotas have been tried out in some cities.

Digital technology offers new opportunities for citizen participation. Clément Mabi (2017, p. 78) analyzes the impact of civic tech: "Digital participation platforms broaden the base of participants and facilitate the collection and analysis of citizen contributions". However, the author warns against the risk of a digital divide and stresses the importance of combining online tools and face-to-face devices.

The question of the articulation between participatory and representative democracy remains a central issue. Yves Sintomer (2020, p. 156) states that "the challenge is to find a balance between the legitimacy of elected representatives and the voice of citizens, without falling into a sterile opposition between these two forms of legitimacy". This requires changes in political and administrative practices, as well as training in participatory methods for elected representatives and civil servants.

Assessing the real impact of participatory arrangements on decision-making remains a major challenge. Guillaume Gourgues (2018, p. 203) points out that "the effect of participatory processes on public policies is often difficult to measure and can vary considerably depending on the context". This observation calls for the development of finer assessment tools and the integration of participation into a broader reflection on the transformation of public action.

Citizen participation also raises questions in terms of the temporality of public action. Hélène Reigner (2017, p. 245) notes that "participatory processes can lengthen decision-making times, which can come into tension with the urgency of certain urban issues". Striking a balance between citizen deliberation time and public action time is a major challenge for decision-makers.

Finally, citizen training and support are essential to ensure informed and constructive participation. Julien Talpin (2019, p. 189) stresses the importance of "citizen empowerment": "Effective participation requires not only spaces for dialogue, but also a strengthening of citizens' capacities to understand the issues and formulate proposals". Initiatives such as popular universities and participatory urban planning workshops can contribute to this objective.

In conclusion, strengthening the mechanisms of citizen participation in decision-making represents a major challenge for contemporary urban governance. As Jacques Donzelot (2020, p. 312) sums up, "the challenge is to move from a democracy of authorization, where citizens delegate their power to elected officials, to a democracy of involvement, where they actively participate in the making of the city". This evolution implies a profound transformation of political and administrative practices, as well as a sustained involvement of citizens in the life of their city.

### 4.2.3 Development of participatory budgets and co-construction of urban projects

Participatory budgets, initiated in Porto Alegre, Brazil, in the 1980s, have gained considerable momentum in France in recent years. Yves Sintomer and Anja Rocke (2019, p. 45) define participatory budgeting as "a mechanism enabling citizens to decide on the allocation of part of the public budget, generally on a municipal scale". This approach aims to strengthen budget transparency and adapt public investments to the needs expressed by local residents.

The implementation of participatory budgets raises several issues. Antoine Bézard (2018, p. 112) points out that "the success of participatory budgets depends largely on the political will of elected officials, the participatory engineering deployed, and the ability to mobilize a broad panel of citizens". The author stresses the importance of effective communication and support for project leaders to guarantee inclusive participation.

The co-construction of urban projects, meanwhile, involves close collaboration between urban planning professionals, elected representatives and residents throughout the design and implementation process. Jodelle Zetlaoui-Léger (2017, p. 78) observes that "this approach challenges the traditional division of roles between experts and laymen, and requires an evolution of professional practices in the field of urban planning". This collaboration can take a variety of forms, from co-design workshops to participatory worksites.

One of the major challenges of co-construction is to articulate the expert knowledge of residents and their knowledge of use. Hélène Hatzfeld (2020, p. 156) asserts that "recognition of citizen expertise is a central challenge for participatory democracy in urban planning". This approach implies developing methods for valorizing and integrating the knowledge gained from residents' everyday experience into the design of urban projects.

Digital technology is playing an increasingly important role in these participatory processes. Fanny Carlet and Éric Hamelin (2016, p. 203) analyze the impact of digital tools on citizen participation in urban planning: "Online platforms for participatory budgeting or urban consultation make it possible to broaden the base of participants and facilitate the collection and analysis of citizen contributions". However, the authors stress the importance of combining these tools with face-to-face arrangements, so as not to exclude populations less at ease with digital technology.

The question of representativeness and inclusion in these participatory processes remains a major issue. Marion Carrel (2018, p. 245) notes that "co-construction schemes often tend to attract already committed citizens and exclude the most marginalized groups". To overcome this bias, initiatives such as drawing lots or "reaching out" to populations far removed from participation have been tried out in some cities.

Assessing the real impact of these approaches on the quality of urban projects and on local democratic life is another crucial point. Loïc Blondiaux and Jean-Michel Fourniau (2019, p. 189) point out that "the effect of participatory processes on urban policies is often difficult to measure and can vary considerably depending on the context". This observation calls for the development of finer assessment tools and the integration of participation into a broader reflection on the transformation of urban public action.

Last but not least, the sustainability and institutionalization of these approaches represent a major challenge. Gilles Pinson (2020, p. 312) observes that "the transition from one-off experiments to a genuine culture of participation in the making of the city requires a profound transformation of institutions and professional practices". In particular, this implies training elected representatives and technicians in participatory methods, as well as a change in the regulatory framework for urban planning.

In conclusion, the development of participatory budgets and the co-construction of urban projects represents a significant evolution in the way cities are conceived and managed. As Jacques Donzelot (2017, p. 367) summarizes, "these participatory approaches aim to reinvent local democracy by making the city not just an object of government, but a political subject in its own right". This ambition implies a sustained commitment from all urban players, from public authorities to citizens, including urban planning professionals and local associations.

## 4.2.4 Inclusion of marginalized groups (women, youth, people with disabilities) in local governance

The notion of inclusion in local governance is part of a broader perspective of social justice and the right to the city. As Marie-Hélène Bacqué (2018, p. 34) points out, "the inclusion of marginalized groups is not limited to simple numerical representation, but implies a transformation of power structures and decision-making processes". This approach aims to go beyond simple consultation to achieve genuine co-construction of urban policies.

The inclusion of women in local governance remains a major challenge. Yves Raibaud (2019, p. 112) observes that "despite legislative advances in parity, women remain under-represented in local decision-making bodies and their specific needs are often overlooked in urban planning". The author argues for a gendered approach to urban planning, taking into account gender-differentiated uses of public space.

The participation of young people in local governance is another crucial issue. Patricia Loncle (2017, p. 78) notes that "involving young people in local decision-making processes can help renew democratic practices and take better account of the specific needs of this age group". Initiatives such as municipal youth councils or participatory budgets dedicated to youth are examples of schemes aimed at fostering this inclusion.

The inclusion of people with disabilities in local governance is part of a logic of universal accessibility. Jean-François Ravaud and Pierre Mormiche (2020, p. 156) emphasize that "the participation of people with disabilities in the design and implementation of urban policies is essential to create truly inclusive cities". This approach involves not only the physical adaptation of the urban environment, but also the accessibility of the participatory processes themselves.

The issue of intersectionality is also crucial in the inclusion of marginalized groups. Éric Fassin (2018, p.

203) highlights that "experiences of marginalization can accumulate and intersect, requiring approaches that take into account the complexity of identities and discriminations". This perspective calls for the development of inclusion schemes that recognize diversity within marginalized groups themselves.

One of the major challenges of inclusion is overcoming barriers to participation. Marion Carrel (2016, p. 245) identifies several obstacles: "mastery of institutional language, availability of time, confidence in oneself and in institutions". To overcome these obstacles, the author recommends "outreach" approaches and training in civic participation.

Digital technology offers new opportunities for the inclusion of marginalized groups. Clément Mabi (2019, p. 189) analyzes the potential of civic tech: "Digital tools can facilitate access to information and participation for certain marginalized groups, but they also risk creating new forms of exclusion linked to the digital divide". The author stresses the importance of combining digital and face-to-face approaches to guarantee effective inclusion.

Assessing the real impact of inclusion policies on local governance remains an important issue. Guillaume Gourgues (2017, p. 312) points out that "the effect of the inclusion of marginalized groups on public policies is often difficult to measure and can vary considerably depending on the context". This observation calls for the development of finer assessment tools and the integration of inclusion into a broader reflection on the transformation of local public action.

Last but not least, sustaining and institutionalizing these inclusion initiatives is a major challenge. Catherine Neveu (2020, p. 367) observes that "the transition from one-off experiments to a genuine culture of inclusion in local governance requires a profound transformation of institutions and mentalities". In particular, this implies training elected representatives and technicians in the challenges of inclusion, as well as changes to the regulatory framework for local democracy.

In conclusion, the inclusion of marginalized groups in local governance represents a crucial challenge for strengthening democracy and equity in our cities. As Jacques Donzelot (2016, p. 415) sums up, "the challenge is to move from a city that is subjugated to a city that is chosen, where every inhabitant, whatever their situation, can help shape their environment". This ambition implies a sustained commitment from all urban players and a constant rethinking of governance practices.

## 4.2.5 Support for neighborhood associations and civic initiatives

The local associative fabric is often seen as the glue that holds neighborhood life together. As Jacques Ion (2017, p. 23) points out, "neighborhood associations play an intermediary role between residents and institutions, while contributing to the creation of social ties and the animation of local life". This mediation function is particularly important in priority urban policy districts, where associations can make up for certain institutional shortcomings.

Support for neighborhood associations takes many forms. Marion Carrel and Catherine Neveu (2018, p. 112) identify several types of support: "financial, logistical, technical, but also in terms of recognition and legitimization". The authors stress the importance of support that preserves the autonomy of associations and their ability to take a critical look at public policies.

Citizen initiatives, meanwhile, reflect a growing desire among residents to get directly involved in transforming their living environment. Agnès Deboulet and Héloïse Nez (2016, p. 78) observe that "these initiatives, whether they be shared gardens, repair cafés or local currencies, embody new forms of citizen engagement, more horizontal and less institutionalized than the traditional associative model". Supporting these initiatives often means changing administrative practices to accommodate their informal and evolving nature.

One of the major challenges in supporting associations and citizen initiatives is to strike a balance between institutional support and maintaining their autonomy. Julien Talpin (2019, p. 156) warns against the risk of instrumentalization: "Overly directive support can lead to a form of domestication of the associative fabric, reducing its capacity for social innovation and counter-power". The author argues for more flexible forms of support, favoring experimentation and mutual learning.

The role of associations and citizens' initiatives in the co-construction of local public policies is increasingly recognized. Hélène Balazard (2020, p. 203) analyzes the emergence of new forms of collaborative governance: "Citizens' councils, neighborhood tables and participatory budgets are all spaces where associations and residents' collectives can contribute directly to the design and implementation of urban policies". This evolution implies a transformation of relations between public authorities and civil society.

The question of the sustainability and professionalization of citizen initiatives raises debate. Gilles Jeannot and Taoufik Souami (2017, p. 245) observe a tension between "the desire to maintain the spontaneous, voluntary nature of these initiatives and the need to ensure their continuity and long-term impact". The authors stress the importance of developing hybrid models, combining civic engagement and professional support.

Digital technology offers new opportunities for the development and networking of local initiatives. Clément Mabi and Anaïs Theviot (2018, p. 189) analyze the impact of civic tech on citizen engagement: "Digital platforms facilitate the coordination of initiatives, the pooling of resources and the visibility of local actions". However, the authors warn against the risk of digital exclusion and stress the importance of combining online tools and physical presence on the ground.

Assessing the social impact of associations and citizen initiatives remains a major challenge. Hélène Rey- Valette et al. (2019, p. 312) stress the need to "develop appropriate evaluation tools, taking into account not only quantifiable results, but also qualitative effects in terms of social ties, empowerment and transformation of practices". This approach implies co-construction of evaluation criteria with the players involved.

Last but not least, training and support for initiative bearers are essential to strengthen the capacity for action of local civil society. Marie-Hélène Bacqué and Mohamed Mechmache (2016, p. 367) advocate "a genuine investment in the power to act of local residents, through training, mentoring and experience-sharing programs". This approach aims to strengthen the skills of local players and foster the emergence of new community leaders.

In conclusion, support for neighborhood associations and citizen initiatives represents an important lever for strengthening local democracy and the resilience of urban communities. As Catherine Neveu (2020, p. 415) sums up, "the challenge is to move from a logic of service provision to a genuine co-production of the city, where citizens are no longer just beneficiaries but full-fledged players in urban transformation". This ambition implies a far-reaching change in institutional practices and greater recognition of the role of civil society in shaping the city.

## 4.3 Urban security, violence and conflict prevention

### 4.3.1 Implementation of situational prevention strategies (lighting, safe urban planning)

Situational prevention is part of a broader approach to urban safety that focuses on the environment rather than the perpetrators. As Sebastian Roché (2018, p. 23) points out, "this approach postulates that delinquency is partly the result of opportunities created by the urban environment, and that it is possible to reduce these opportunities through appropriate planning". This perspective marks a paradigm shift from purely repressive or social approaches to safety.

Urban lighting is one of the central elements of situational prevention. Jean-Michel Deleuil (2019, p. 112) analyzes the evolution of street lighting policies: "Beyond its primary function of nighttime visibility, lighting is

increasingly conceived as a tool for making public spaces safer, aimed at reducing feelings of insecurity and deterring deviant behavior". The author notes, however, that the actual effectiveness of lighting in reducing crime is the subject of scientific debate.

Safe urban design encompasses a wide range of measures. Virginie Malochet (2017, p. 78) identifies several recurring principles: "legibility of spaces, natural control of access, territoriality, environmental quality and place management". These principles are reflected in concrete developments such as the elimination of nooks and crannies, the installation of anti-grouping street furniture, or the creation of transitional spaces between the public and the private.

One of the major challenges of situational prevention is to reconcile safety with the quality of urban life. Alain Bauer and Christophe Soullez (2020, p. 156) warn against the risk of creating unfriendly "defensive spaces": "Planning that focuses too much on security can lead to a form of bunkerization of public space, to the detriment of its inclusive character and social vitality". The authors call for a balanced approach, integrating safety issues into a broader reflection on the quality of public space.

The actual effectiveness of situational prevention strategies is a matter of debate. Tanguy Le Goff and Jacques de Maillard (2016, p. 203) highlight the difficulty of assessing the impact of these measures: "The effects of situational prevention are often diffuse and difficult to isolate from other factors influencing urban security". The authors also note the risk of crime shifting to less secure areas.

The situational approach also raises ethical and political questions. Frédéric Ocqueteau (2018, p. 245) criticizes a potentially reductive vision of security: "By focusing on the physical environment, this approach risks neglecting the social and economic causes of delinquency, and promoting a technicist vision of security to the detriment of a more social and preventive approach". This criticism calls for situational prevention to be integrated into a more comprehensive urban social development strategy.

Involving residents in the design of safety features is increasingly seen as essential. Julie Sedel (2019, p. 189) analyzes the emergence of participatory approaches to urban safety: "Involving users in defining safety problems and designing solutions not only makes it possible to adapt developments to real needs, but also to strengthen the sense of ownership of public space". This participatory approach can help to overcome a purely technocratic vision of safety.

Finally, the integration of new technologies into situational prevention raises new issues. Eric Heilmann (2017, p. 312) analyzes the impact of video surveillance and detection technologies: "These tools offer new possibilities in terms of surveillance and responsiveness, but also raise questions in terms of privacy and social control". The author calls for ethical reflection on the use of these technologies in public spaces.

In conclusion, the implementation of situational prevention strategies represents a significant evolution in the approach to urban safety. As Maurice Cusson

(2020, p. 415) sums up, "the challenge is to create urban environments that are both safe and liveable, where security is not synonymous with closure and exclusion". This ambition implies an interdisciplinary approach, combining expertise in urban planning, safety and social sciences, as well as the active involvement of residents in the co-construction of their living environment.

## 4.3.2 Development of social prevention programs targeting at-risk youth

Social prevention with young people is part of a comprehensive approach to safety that goes beyond mere repression. As Laurent Mucchielli (2018, p. 23) points out, "this approach recognizes that juvenile delinquency is often a symptom of deeper social problems, requiring multidimensional interventions". This perspective implies close collaboration between security, education, health and social work players.

Identifying "at-risk" young people is a crucial but delicate stage in these programs. Sébastien Roché (2017, p. 112) warns against the risks of stigmatization: "Targeting interventions at certain young people can lead to a form of self-fulfilling prophecy, reinforcing their sense of marginalization". The author argues for a more global approach, focusing on at-risk environments rather than individuals.

Social prevention programs take a variety of forms. Marwan Mohammed (2019, p. 78) identifies several recurring areas of intervention: "support for parenthood, reinforcement of psychosocial skills, school support, professional integration and structured leisure activities". These interventions aim to reinforce protective factors and reduce risk factors in young people's environment.

One of the major challenges of these programs is the involvement of the young people themselves. Patricia Loncle (2016, p. 156) stresses the importance of a participatory approach: "The most effective interventions are those that actively involve young people in defining their needs and designing solutions". This approach not only enables interventions to be adapted to the realities experienced by young people, but also reinforces their sense of empowerment.

The question of evaluating the effectiveness of social prevention programs remains a major challenge. Sebastian Roché and Jacques de Maillard (2020, p. 203) note the difficulty of measuring the long-term effects of these interventions: "The impacts of social prevention are often diffuse and manifest themselves over the long term, which complicates their evaluation". The authors call for the development of more sophisticated evaluation tools, combining quantitative and qualitative approaches.

The training and professionalization of streetworkers play a crucial role in the success of these programs. Véronique Le Goaziou (2018, p. 245) analyzes the evolution of the profession of street educator: "Faced with the complexity of the situations they encounter, street educators have to develop multiple skills, combining social work, mediation and in-depth knowledge of territorial realities". This evolution implies significant investment in ongoing training and supervision for professionals.

Coordination between the various players involved in social prevention is another major challenge. Michel Marcus (2017, p. 189) stresses the importance of local partnerships: "The effectiveness of interventions depends on close coordination between schools, social services, the police, the justice system and local associations". This partnership approach often requires the establishment of specific coordination structures, such as Local Councils for Security and Delinquency Prevention (Conseils Locaux de Sécurité et de Prévention de la Délinquance - CLSPD).

The relationship between prevention and repression remains a subject of debate. Alain Bauer and Christophe Soullez (2019, p. 312) argue for a balanced approach: "An effective juvenile delinquency prevention policy must combine early intervention and a clear normative framework, with graduated responses in the event of delinquency". This approach implies close collaboration between those involved in prevention and those involved in juvenile justice.

Last but not least, it's essential to take into account specific territorial features when developing these programs. Joëlle Bordet (2016, p. 367) stresses the importance of adapting interventions to local realities: "Prevention programs must take account of the social, cultural and economic dynamics specific to each territory, drawing on local resources and initiatives". This territorial approach implies detailed knowledge of neighborhoods and the ability to mobilize local players.

In conclusion, the development of social prevention programs targeting at-risk youth represents a complex but essential challenge for social cohesion and urban safety. As François Dubet (2020, p. 415) sums up, "the challenge is to build interventions that enable young people to project themselves positively in society, by giving them the means to develop their potential and embark on pathways to success". This ambition implies a global, interdisciplinary approach rooted in local realities, mobilizing all the players involved in youth and urban social development.

### 4.3.3 Strengthening community policing and local security partnerships

Strengthening community policing and local security partnerships is part of a comprehensive approach to preventing and combating urban insecurity. This strategy aims to establish a relationship of trust between law enforcement agencies and the population, while involving various local players in the co-production of security.

Proximity policing, a concept developed in France in the 1990s, is based on a greater presence of officers in the field and a better understanding of local realities. According to Sebastian Roché (2005, p. 78), "community policing aims to bring the police closer to the population by encouraging contact, listening and problem-solving at local level". This approach allows for more preventive intervention, adapted to the specific needs of each neighborhood.

Local security partnerships, on the other hand, are based on the mobilization of a network of diverse players: law enforcement agencies, local elected representatives, social landlords, associations, shopkeepers and so on. As Tanguy Le Goff (2008, p. 124) points out, "these partnerships make it possible to pool resources and skills for a more effective, coordinated approach to safety issues". They also encourage a better flow of information and greater responsiveness to problem situations.

Implementing these systems requires a reorganization of police services and appropriate training for officers. According to Jacques de Maillard and Mathieu Zagrodzki (2017, p. 56), "the success of community policing relies on a change in professional culture, moving from a repressive logic to one of public service and problem-solving". This involves developing skills in mediation, communication and analysis of local issues.

Evaluating these schemes remains a complex task, but several studies have shown encouraging results. Virginie Malochet (2009, p. 201) notes that "community policing and local security partnerships help to improve residents' sense of security and reduce certain forms of delinquency, particularly incivilities and public disorder".

However, these approaches are also subject to criticism. Some authors, such as Frédéric Ocqueteau (2004, p. 89), point to "the risk of a dilution of responsibilities and a loss of operational efficiency on the part of law enforcement agencies". It is therefore crucial to strike a balance between proximity and maintaining intervention capabilities.

In conclusion, the strengthening of community policing and local security partnerships represents a major evolution in the governance of urban security. By fostering a more collaborative approach rooted in local realities, these measures offer promising prospects for improving safety and quality of life in French cities. Nevertheless, their implementation remains a challenge that requires a sustained commitment from public authorities and constant adaptation to changing urban issues.

### 4.3.4 Fighting organized crime and illicit trafficking in urban areas

The fight against organized crime and illicit trafficking in urban areas represents a major challenge for the French authorities, requiring a multidimensional and coordinated approach. This complex phenomenon is often rooted in sensitive urban areas, where precarious socio-economic conditions can encourage the development of illegal activities.

According to Nacer Lalam (2018, p. 45), "organized crime in urban areas is characterized by its ability to adapt to social and technological changes, making it particularly difficult to eradicate". Criminal networks exploit urban vulnerabilities, such as population density, relative anonymity and the presence of transport infrastructures, to develop their illicit activities.

Drug trafficking is one of the main manifestations of organized urban crime. As Michel Kokoreff (2010, p. 78) points out, "drug trafficking has become deeply entrenched in certain neighborhoods, creating a parallel economy that undermines the local social and economic fabric". This situation calls for a combination of law enforcement and preventive action.

To effectively combat these phenomena, the French authorities have implemented a variety of strategies. Laurent Mucchielli (2014, p. 132) notes that "the traditional police approach has been supplemented by more sophisticated intelligence devices and enhanced international cooperation". The Groupes Interministériels de Recherches (GIR) illustrate this evolution, combining the skills of different government departments to target criminal assets.

At the same time, prevention plays a crucial role. According to Marwan Mohammed (2012, p. 201), "prevention programs targeting at-risk young people in sensitive neighborhoods can help reduce recruitment by criminal networks". These initiatives often include vocational integration and social support.

Urban and regional planning also play a part in this fight. As explained by Sébastian Roché (2020, p. 67), "urban renewal and social diversity can help reduce the number of lawless areas conducive to the development of criminal activities". This approach aims to transform the physical and social environment to make it less favorable to illicit activities.

However, despite these efforts, challenges persist. Fabien Jobard (2015, p. 189) points out that "the growing sophistication of criminal networks, particularly in their use of digital technologies, requires constant adaptation of combat strategies". What's more, the transnational dimension of certain types of trafficking complicates the action of national authorities.

In conclusion, the fight against organized crime and illicit trafficking in urban areas in France requires a comprehensive approach, combining law enforcement, prevention and urban transformation. While progress has been made, the constant evolution of criminal phenomena demands constant vigilance and adaptation of public policies. The involvement of all local, national and international players remains crucial if we are to achieve lasting results in this complex field.

## 4.3.5 Managing conflicts over the use of public space and urban mediation

Managing conflicts over the use of public space and urban mediation have become major issues in the development of modern cities, where densification and diversification of populations are generating growing tensions. These conflicts can take many forms, from the exclusive appropriation of certain spaces by specific groups to noise nuisance, and disagreements over the use of public facilities (Blanc, Maurice, 2006, p. 27).

In response to these challenges, many cities have set up urban mediation schemes. According to Bonafé-Schmitt, Jean-Pierre (2010, p. 45), urban mediation can be defined as "a process for creating and repairing social ties and resolving everyday conflicts, in which an impartial and independent third party attempts, through the organization of exchanges between people or institutions, to help them improve a relationship or settle a conflict between them".

The effectiveness of urban mediation depends on several key factors. Firstly, it is essential to train professional mediators capable of understanding the complex social dynamics at work in urban space. These mediators must be able to identify the issues underlying apparent conflicts and facilitate dialogue between stakeholders (Stébé, Jean-Marc, 2005, p. 112).

In addition, the introduction of preventive consultation mechanisms can help to anticipate and defuse potential conflicts. As Blondiaux, Loïc (2008, p. 78) points out, "involving residents in the design and management of public spaces means that their needs and expectations are better taken into account, thus reducing the risk of subsequent conflicts".

Managing conflicts over the use of public space also requires a multidimensional approach, integrating urban planning, social and regulatory aspects. According to Jacquier, Claude (2011, p. 203), "the layout of public space must be designed to promote the peaceful coexistence of different uses and users, by creating buffer zones, diversifying functions and providing transitional spaces".

The use of new technologies can also play an important role in preventing and resolving conflicts of use. Mobile applications for reporting problems or participating in online consultations are increasingly used to facilitate communication between citizens and local authorities (Cardon, Dominique, 2015, p. 167).

Finally, it is crucial to recognize that managing conflicts over the use of public space is part of a broader perspective of social cohesion and living together. As Donzelot, Jacques (2012, p. 89) reminds us, "the inclusive city cannot be decreed; it is constructed on a daily basis through processes of negotiation and collective learning about how to share urban space".

In conclusion, effective management of conflicts over the use of public space and urban mediation require an integrated approach, combining mediation skills, urban planning that is sensitive to social issues, mechanisms for citizen participation and a strong political will to promote living together in diversity. These efforts contribute not only to reducing tensions in urban spaces, but also to strengthening social ties and building more resilient and inclusive cities.

## 4.4 Cultural heritage, local identity and promotion of specific local features

### 4.4.1 Preserving and restoring architectural and urban heritage

The preservation and rehabilitation of architectural and urban heritage are crucial issues for contemporary cities, combining the preservation of historical identity with the need to adapt to modern needs. This approach is part of a sustainable development perspective, making the most of existing resources while meeting current requirements in terms of housing and urban planning.

According to Choay, Françoise (2009, p. 32), "architectural and urban heritage represents not only a cultural legacy, but also an economic and social capital whose enhancement can contribute significantly to local development". This approach underlines the importance of considering heritage as a dynamic asset rather than a mere relic of the past to be preserved.

Preserving architectural heritage often involves complex technical challenges. As Pérouse de Montclos, Jean-Marie (2011, p. 156) notes, "the restoration of historic buildings requires specialized expertise and a thorough understanding of ancient construction techniques, in order to respect the integrity of the work while ensuring its durability". This requirement implies the training of specialized professionals and the development of appropriate intervention methods.

Furthermore, the rehabilitation of urban heritage is part of a broader approach to urban renewal. According to Gravari-Barbas, Maria (2013, p. 89), "the revitalization of historic centers can act as a catalyst for urban regeneration, attracting new residents and stimulating economic activity". However, this dynamic also raises issues of social equity, including the risk of gentrification and exclusion of local populations.

The integration of new technologies in heritage preservation offers promising prospects. Gadaud, Christophe (2018, p. 204) points out that "the use of 3D modeling and augmented reality makes it possible not only to document heritage accurately, but also to make it more accessible and understandable for the general public".

The participatory dimension is also playing a growing role in preservation and rehabilitation projects. As Bacqué, Marie-Hélène (2015, p. 127) asserts, "involving residents in the definition and implementation of rehabilitation projects makes it possible to take better account of local uses and expectations, while strengthening the community's ownership of its heritage".

Finally, the preservation of architectural and urban heritage must be part of a long-term vision of urban development. According to Veschambre, Vincent (2008, p. 73), "heritage preservation should not be seen as an end in itself, but as a lever for building more sustainable and inclusive cities, by combining respect for heritage with innovation".

In conclusion, the preservation and rehabilitation of architectural and urban heritage represents a multi-dimensional challenge, requiring an integrated approach that combines technical expertise, cultural sensitivity, citizen participation and strategic vision. When properly implemented, this approach can make a significant contribution to the attractiveness of cities, the quality of life of residents and the construction of a strong, shared urban identity.

## 4.4.2 Promoting local cultural traditions and craft skills

Promoting local cultural traditions and craft skills is part of an approach to preserving and enhancing intangible heritage, which is essential to the identity and dynamism of urban areas. This approach aims to keep alive cultural practices inherited from the past, while adapting them to contemporary realities.

According to Chevallier, Denis (2012, p. 45), "cultural traditions and craft skills constitute invaluable cultural and social capital, conveying meaning and identity for local communities". This perspective underlines the importance of these practices not only as historical heritage, but also as a vector of social cohesion and territorial anchorage.

The transmission of craft skills represents a major challenge in the face of modernization and industrialization. As Choay, Françoise (2015, p. 112) notes, "the preservation of traditional craft techniques should not be seen as resistance to progress, but as an enriching complement to contemporary modes of production". This approach implies the implementation of appropriate training and apprenticeship schemes, enabling the intergenerational transmission of skills.

Integrating local cultural traditions into urban development can contribute significantly to the attractiveness of territories. Gravari-Barbas, Maria and Violier, Philippe (2017, p. 78) point out that "enhancing local cultural specificities helps to differentiate cities in a context of heightened territorial competition, while reinforcing inhabitants' sense of belonging".

The development of cultural tourism offers opportunities for the promotion of local traditions and crafts. However, as Poulot, Dominique (2014, p. 203) warns, "it is crucial to strike a balance between tourism enhancement and the preservation of the authenticity of cultural practices, in order to avoid their folklorization or excessive commodification".

Innovation also plays an important role in revitalizing traditional know-how. According to Fraysse, Patrick (2016, p. 89), "combining craft techniques with new technologies can open up new creative perspectives, helping to renew interest in these practices among younger generations".

The economic dimension should not be overlooked when promoting cultural traditions and craftsmanship. As Greffe, Xavier (2013, p. 156) asserts, "the development of craft industries and the creative economy based on local know-how can be a significant lever for economic development and the creation of jobs that cannot be relocated".

Finally, the promotion of local cultural traditions is part of a broader reflection on cultural diversity in urban environments. According to Amselle, Jean-Loup (2011, p. 67), "the valorization of traditional cultural practices must be accompanied by an openness to interculturality and exchange, in order to avoid identitarian withdrawal and encourage dialogue between different urban communities".

In conclusion, the promotion of local cultural traditions and craft skills represents a complex challenge, combining heritage preservation, economic development, social cohesion and innovation. This requires an integrated approach, involving public, private and associative players, to ensure the durability and dynamism of these practices in the contemporary urban context. It thus contributes to shaping more authentic, creative and resilient cities, rooted in their history yet open to the future.

### 4.4.3 Development of local cultural facilities (libraries, cultural centers)

The development of local cultural facilities, such as libraries and cultural centers, plays a crucial role in democratizing access to culture and strengthening the urban social fabric. These structures are essential anchors for local cultural life and make a significant contribution to improving the quality of life in neighborhoods.

According to Saez, Guy (2014, p. 78), "local cultural facilities represent essential public spaces, fostering social mixing and the exchange of knowledge within urban communities". This perspective underscores the importance of these venues not only as vectors of cultural dissemination, but also as catalysts of social ties.

Libraries, in particular, are undergoing a major evolution in their role and functions. As Bertrand, Anne-Marie (2011, p. 123) notes, "modern libraries are transforming themselves into veritable centers of social and cultural life, going beyond their traditional function of lending books to become places of learning, creation and encounter". This transformation involves redefining the layout of spaces and diversifying the services on offer.

The accessibility of local cultural facilities is a major issue in ensuring their effectiveness. According to Fleury, Antoine (2017, p. 56), "the balanced spatial distribution of cultural facilities within the urban fabric is essential to reduce inequalities in access to culture and encourage the participation of all inhabitants". This approach requires careful urban planning and coordination between cultural and land-use policies.

The development of multi-purpose cultural centers meets a growing demand for accessible venues for artistic expression and creation. As Lextrait, Fabrice (2016, p. 201) asserts, "new cultural spaces must be designed as hybrid venues, capable of hosting a diversity of artistic practices and fostering cross-fertilization between amateurs and professionals".

The integration of new technologies into local cultural facilities is opening up new perspectives. According to Paquienséguy, Françoise (2018, p. 89), "the digitization of resources and the offer of online services make it possible to considerably extend the reach of cultural facilities, while at the same time responding to audiences' new modes of cultural consumption".

The participation of local residents in the design and management of local cultural facilities is increasingly encouraged. As Bacqué, Marie-Hélène and Mechmache, Mohamed (2013, p. 45) point out, "involving citizens in the definition of local cultural projects makes it possible to better respond to the specific needs of communities and strengthen the appropriation of these spaces by residents".

Finally, the development of local cultural facilities is part of a broader reflection on the place of culture in sustainable urban development. According to Lefebvre, Alain (2012, p. 167), "local cultural policies must be conceived in articulation with the social, economic and environmental challenges of the territory, in order to contribute fully to the quality of urban life".

In conclusion, the development of local cultural facilities represents an essential investment in the cultural and social vitality of towns and cities. When designed to be inclusive and adapted to local needs, these structures can play a major role in reducing cultural inequalities, strengthening social cohesion and stimulating urban creativity. Their implementation requires a cross-disciplinary approach, involving close collaboration between cultural players, urban planners, local elected representatives and citizens, to create lively, accessible cultural spaces rooted in the reality of their territory.

## 4.4.4 Support for contemporary artistic creation and urban cultural expression

Supporting contemporary artistic creation and urban cultural expression is an essential part of the drive for cultural vitality and innovation in urban spaces. This approach aims to encourage the emergence of new artistic forms, while promoting the specific cultural characteristics of the urban context.

According to Heinich, Nathalie (2014, p. 78), "contemporary artistic creation plays a crucial role in the renewal of collective representations and in the ability of societies to think about themselves". This perspective underlines the importance of contemporary art not only as an aesthetic expression, but also as a vehicle for social and political reflection.

Urban cultural expressions such as street art, street art and urban music are playing an increasingly important role in the cultural landscape of cities. As Vivant, Elsa (2009, p. 123) notes, "these emerging artistic forms are contributing to the reappropriation of public space by citizens and the redefinition of urban identities". This dynamic implies greater institutional recognition of these long-marginalized practices.

Support for contemporary artistic creation often involves the provision of working and creative spaces. According to Grésillon, Boris (2016, p. 201), "artistic wastelands and cultural third places play a crucial role in the urban creative ecosystem, offering spaces for experimentation and collaboration conducive to artistic innovation". These alternative spaces also play a part in the urban regeneration of changing neighborhoods.

The integration of contemporary art into public space raises important issues in terms of urban planning. As Ruby, Christian (2012, p. 56) states, "contemporary public art must not be limited to a decorative function, but must engage in a dialogue with the urban environment and its inhabitants". This approach requires close collaboration between artists, urban planners and public decision-makers.

The development of urban festivals and cultural events offers significant opportunities for the promotion of contemporary creation. According to Gravari- Barbas, Maria and Veschambre, Vincent (2015, p. 89), "temporary cultural events make it possible to experiment with new forms of artistic expression and reach diversified audiences, while contributing to the attractiveness of territories".

Support for urban cultural expression also

requires a rethink of the ways in which creative work is financed and economically valorized. As Greffe, Xavier (2010, p. 145) points out, "the development of the urban creative economy requires appropriate support mechanisms, combining public funding, private patronage and new forms of participatory financing".

The participatory dimension is playing a growing role in contemporary artistic creation in urban environments. According to Zask, Joëlle (2013, p. 167), "involving local residents in artistic creation processes not only democratizes access to art, but also generates works rooted in the social and cultural reality of the territories".

Finally, support for contemporary artistic creation and urban cultural expressions is part of a broader reflection on cultural diversity and interculturality in the urban environment. As Amselle, Jean-Loup (2017, p. 92) states, "the promotion of hybrid and transcultural artistic forms can contribute to the construction of plural urban identities that are open to the world".

In conclusion, support for contemporary artistic creation and urban cultural expression represents a major challenge for cultural vitality and innovation in contemporary cities. This requires a multi-dimensional approach, combining ambitious cultural policies, urban planning that is sensitive to artistic creation, and citizen involvement in creative processes. By encouraging the emergence of new forms of expression and enhancing urban cultural diversity, these initiatives help to shape more creative, inclusive and resilient cities, capable of responding to contemporary challenges through art and culture.

## 4.4.5 Enhancing local identity in urban development projects

Enhancing local identity in urban development projects has become a crucial issue for many cities seeking to preserve their unique character while adapting to contemporary challenges. This approach is part of a broader vision of urban planning that recognizes the importance of cultural and historical context in the design of urban spaces. According to Françoise Choay (2015, p. 78), "urban identity is not simply the preservation of historic buildings, but encompasses all the social practices, traditions and collective representations that give meaning to a place". This holistic perspective implies careful consideration of how development projects can not only respect but also reinforce the existing social and cultural fabric.

The integration of local identity into urban planning is achieved through a variety of strategies. One of these is to incorporate traditional architectural elements or materials into new construction, creating a dialogue between past and present. Philippe Panerai (2018, p. 132) points out that "this approach maintains visual and symbolic continuity with the history of the place, while meeting contemporary functional needs".

However, he warns against a superficial or pastiche use of these elements, which would risk distorting their original meaning.

Another important aspect of local identity enhancement is the preservation and reinterpretation of traditional public spaces. Market squares, historic promenades or community gathering places play a crucial role in the social life and collective identity of cities. Jean-Pierre Charbonneau (2017, p. 215) asserts that "these spaces are the true guardians of urban memory and must be at the heart of redevelopment projects". Their revitalization can involve a subtle modernization of infrastructure while preserving their original social and symbolic function.

Citizen participation also plays a central role in enhancing local identity. By involving residents in the design and decision-making process, urban planners can better understand and integrate community values and aspirations. Marie-Hélène Bacqué and Mario Gauthier (2016, p. 45) argue that "the co-construction of urban projects not only guarantees their social acceptability, but also enriches their content with the contribution of local knowledge". This participatory approach can take a variety of forms, from public consultations to collaborative design workshops.

Nevertheless, the enhancement of local identity in urban planning also raises challenges. One of these is the risk of "museumizing" certain neighborhoods, which could lead to their gentrification and the exclusion of local populations. Vincent Veschambre (2019, p. 167) warns against "excessive patrimonialization, which freezes the city in an idealized image of its past, to the detriment of its social and economic vitality". It is therefore crucial to strike a balance between preservation and evolution, ensuring that development projects serve the real needs of today's inhabitants while respecting the historical legacy.

In conclusion, enhancing local identity in urban development projects represents a complex but essential challenge for creating liveable, sustainable and culturally rich cities. It requires a multidisciplinary approach, combining urban planning expertise, historical sensitivity and community involvement. As Michel Lussault (2020, p. 298) sums up, "the challenge is to build cities that are not mere functional aggregates, but meaningful places where each inhabitant can recognize himself and flourish". This holistic vision of urban development paves the way for more resilient, inclusive and authentic cities.

## Conclusion:

Social resilience and the fight against urban inequalities are essential pillars of sustainable urban development, encompassing the many interconnected dimensions of city life. Social cohesion in urban areas can only be achieved through an integrated approach, combining housing, education, health and civic participation policies.

Improving access to basic services and reducing urban poverty are the cornerstones of this approach. The fight against urban inequalities necessarily involves strengthening social infrastructures and a proactive policy of inclusion. This approach implies significant investment in social housing, health and education facilities, as well as in support programs for vulnerable populations.

Promoting social cohesion and citizen participation is emerging as a powerful lever for urban transformation. The involvement of citizens in local governance is not limited to a consultative dimension, but must extend to a genuine co-construction of urban policies. This participatory dynamic helps to reinforce a sense of belonging and legitimize public decisions.

Urban safety and violence prevention are part of a comprehensive approach to urban quality of life. Situational and social prevention strategies, combined with the strengthening of community policing, help to create a safer, more inclusive urban environment. Conflict management and urban mediation also play a crucial role in maintaining social harmony.

Finally, enhancing cultural heritage and local identity is an essential element of urban social resilience. Preserving architectural heritage, promoting cultural traditions and supporting contemporary artistic creation contribute not only to the attractiveness of cities, but also to strengthening social ties and affirming plural urban identities.

In conclusion, building resilient and equitable cities requires a holistic approach, integrating ambitious social policies, inclusive governance and the enhancement of local cultural resources. This implies close collaboration between public and private players and civil society, as well as a long-term vision of urban development. By meeting these challenges, cities can become fairer, more dynamic and more sustainable places to live, capable of adapting to the social, economic and environmental changes of the 21st century.

# *Chapter 5: Environmental resilience and adaptation to climate change*

Environmental resilience and adaptation to climate change have become crucial issues for cities in the 21st century. Faced with the acceleration of global warming and the degradation of ecosystems, urban areas, which concentrate a growing proportion of the world's population, are particularly vulnerable to environmental impacts. This vulnerability is exacerbated by demographic density, the concentration of economic activities and the complexity of urban infrastructures. In this context, it is imperative to rethink our urban development models to make them more sustainable and resilient.

Sustainable management of natural resources and protection of ecosystems are the cornerstones of this urban ecological transition. This involves preserving and restoring green spaces, managing water resources in an integrated way, improving air quality, promoting efficient waste management and protecting urban and peri-urban biodiversity. These actions aim to maintain the ecological balance of cities while improving the quality of life of their inhabitants.

Energy efficiency and the development of renewable energies are also at the heart of urban adaptation strategies. Reducing the energy consumption of buildings, deploying local renewable energy production systems, introducing sustainable urban transport and supporting eco-industries all contribute to decarbonizing the urban economy. These initiatives are accompanied by awareness-raising and environmental education efforts to encourage citizens to adopt eco-responsible behavior.

Resilient urban planning and regional development play a crucial role in adapting cities to climate change. Integrating climate issues into urban planning documents, developing eco-friendly neighborhoods, adapting urban infrastructures and managing urban heat islands are all levers for creating cities that are more resilient to climatic hazards. Promoting sustainable urban and peri-urban agriculture also helps to strengthen food security and the resilience of urban systems.

Finally, reducing the risk of natural disasters and setting up effective warning systems are essential to protect urban populations from extreme climatic events. This involves accurate mapping of at-risk areas, reinforcing critical infrastructures, developing early warning systems and strengthening local crisis management capacities. Regional and international cooperation plays an important role in sharing knowledge and resources to improve the climate resilience of cities.

This holistic approach to urban environmental resilience requires close coordination between the various urban players, long-term planning and substantial investment. It represents a major challenge for 21st-century cities, but also an opportunity to fundamentally rethink our urban lifestyles to make them more sustainable, inclusive and resilient in the face of future environmental challenges.

## 5.1 Sustainable management of natural resources and protection of ecosystems

### 5.1.1 Preserving and restoring urban green spaces

The preservation and restoration of urban green spaces is a fundamental pillar of environmental resilience in today's cities. These spaces play a crucial role in improving urban quality of life, regulating the local microclimate and preserving biodiversity. According to a study by Catherine Larrère and Raphaël Larrère (2015, p. 78), urban green spaces make a significant contribution to reducing urban heat islands, improving air quality and managing stormwater, while offering places for relaxation and recreation that are essential to the well-being of city dwellers.

Preserving existing green spaces requires a multi-dimensional approach, integrating legal, urban planning and ecological aspects. Philippe Clergeau (2020, p. 145) stresses the importance of safeguarding these spaces in urban planning documents, while implementing ecological management measures to maintain their biodiversity and ecosystem functions. This preservation also involves combating soil artificialisation and limiting urban sprawl, as advocated by Nathalie Blanc and Cyria Emelianoff (2019, p. 203) in their work on sustainable urbanism.

Restoring degraded green spaces or creating new ones in dense urban fabric represents a major challenge for many cities. Emmanuelle Baudry and Audrey Muratet (2018, p. 56) highlight the importance of recreating ecological continuities within cities, through the establishment of green and blue grids. These ecological corridors link different green spaces together, encouraging the circulation of species and maintaining urban biodiversity.

Involving citizens in the management and maintenance of urban green spaces is also a crucial aspect of their preservation and restoration. The work of Sandrine Manusset (2017, p. 112) shows that participatory urban gardening initiatives or the creation of shared gardens strengthen social ties and the appropriation of green spaces by residents, while contributing to environmental awareness.

Valuing the ecosystem services provided by urban green spaces is a powerful argument for justifying investment in their preservation and restoration. Luc Abbadie and Yves Lestienne (2021, p. 89) have quantified these services in economic terms, showing that the benefits in terms of public health, climate regulation and quality of life far outweigh the costs of maintaining and managing these spaces.

Finally, innovation in urban greening techniques is opening up new prospects for restoring and creating green spaces in cities. Research by Frédéric Madre and Philippe Clergeau (2016, p. 234) on vegetated roofs and facades shows the potential of these solutions to increase the vegetated surface area in dense urban environments, while providing benefits in terms of thermal insulation and stormwater management.

In conclusion, preserving and restoring urban green spaces requires an integrated approach, combining urban planning, ecological management, citizen participation and technological innovation. These efforts are essential to building more resilient cities in the face of the environmental and climatic challenges of the 21st century.

## 5.1.2 Integrated water resource management and combating water stress

Integrated water resource management and the fight against water stress have become major challenges for 21st century cities, faced with increasing pressure on their water resources. This complex issue calls for a holistic approach, taking into account the environmental, social, economic and technical aspects of urban water management.

Bernard Barraqué and Pierre-Alain Roche (2020, p. 67) emphasize the importance of integrated water resource management on a watershed scale, transcending traditional administrative boundaries. This approach enables better account to be taken of the interactions between different water uses (domestic, industrial, agricultural) and aquatic ecosystems. The authors stress the need to strengthen water governance through multi-stakeholder consultation structures, such as the Basin Committees in France.

Combating water stress in urban areas involves diversifying water supply sources. Ghislain de Marsily (2018, p. 123) highlights the importance of developing alternative solutions such as the reuse of treated wastewater or rainwater harvesting. These practices reduce pressure on conventional water resources and increase the resilience of urban supply systems to climate hazards.

Improving the efficiency of drinking water distribution networks is also an important lever in the fight against water stress. Jean-Claude Deutsch and Isabelle Gautheron (2019, p. 201) highlight the importance of reducing network leakage, which can account for up to 20% of distributed volumes in some French cities. The authors advocate the use of intelligent technologies for the rapid detection and repair of leaks, as well as the implementation of programs to renew aging infrastructures.

Water demand management is another crucial aspect of the fight against water stress. The work of Marielle Montginoul (2017, p. 89) highlights the effectiveness of pricing incentives and awareness campaigns in reducing household water consumption. The author also stresses the importance of adapting these measures to local socio-economic contexts to ensure their social acceptability.

Preserving and restoring urban and peri-urban aquatic ecosystems plays an essential role in integrated water resource management. Jean-Louis Rivière and Cécile Clavel (2021, p. 156) demonstrate that urban wetlands, watercourses and green spaces contribute to natural water purification, groundwater recharge and flood regulation. The authors call for these spaces to be better integrated into urban planning and stormwater management.

Adapting to climate change is a major challenge for urban water resource management. Emma Haziza and Jean Jouzel (2020, p. 234) emphasize the need to anticipate the impacts of climate change on the local water cycle. They advocate the development of hydrological models that take climate scenarios into account, in order to adapt long-term water management strategies.

Finally, technological innovation offers new prospects for more efficient management of water resources. Bruno Tassin and Jean-Marie Mouchel (2018, p. 178) highlight the potential of information and communication technologies to optimize water network management, improve consumption forecasting and facilitate citizen participation in water management.

In conclusion, integrated water resource management and the fight against water stress in urban environments require a multidimensional approach, combining technical, organizational and behavioral solutions. This approach must be part of a long-term vision, taking into account the challenges of climate change and the preservation of aquatic ecosystems.

### 5.1.3 Improving air quality and reducing air pollution

Improving air quality and reducing air pollution are major challenges for public health and the environment in urban areas. This complex issue calls for a multidisciplinary approach and concerted action at different levels.

Isabella Annesi-Maesano and Rémy Slama (2019, p. 45) highlight the considerable impact of air pollution on the health of urban populations. Their work highlights the links between chronic exposure to fine particles and the increase in respiratory, cardiovascular and certain cancers. The authors stress the need for ambitious emission reduction policies to protect public health.

Reducing road traffic emissions is a major lever for improving urban air quality. Mathieu Saujot and Laura Brimont (2021, p. 112) analyze the effectiveness of low-emission zones (ZFE) introduced in several French cities. Their study shows that these schemes, coupled with support measures for the transition to less polluting vehicles, can significantly reduce concentrations of nitrogen dioxide and fine particles in urban air.

Urban planning also plays a crucial role in improving air quality. According to Sébastien Bourdin and André Torre (2020, p. 78), the design of compact cities that encourage soft mobility and public transport can reduce travel-related emissions. The authors emphasize the importance of integrating air quality issues right from the design phase of urban projects, notably through the creation of natural ventilation corridors.

Improving the energy efficiency of buildings is another important area for reducing air pollution. Marie-Hélène Laurent and Bruno Peuportier (2018, p. 156) demonstrate that thermal renovation of the building stock and the development of renewable energies for district heating can considerably reduce emissions of atmospheric pollutants linked to the residential and tertiary sectors.

Air quality monitoring and forecasting are essential for informing and protecting populations. Augustin Colette and Laurence Rouïl (2017, p. 203) present recent advances in air quality forecasting models, integrating satellite data and connected urban sensors. These tools make it possible to anticipate pollution peaks and implement appropriate preventive measures.

Urban greening is increasingly recognized as a complementary solution for improving air quality. Philippe Clergeau and Nathalie Blanc (2022, p. 89) highlight the role of trees and urban green spaces in filtering atmospheric pollutants and regulating the urban microclimate. The authors stress, however, the need to select suitable plant species to maximize air quality benefits while limiting emissions of biogenic volatile organic compounds.

Involving citizens in the fight against air pollution is also an important aspect. Isabelle Roussel and Lionel Charles (2020, p. 234) analyze the impact of participatory science initiatives on public awareness and commitment to improving air quality. Their work shows that these initiatives contribute to a better understanding of the issues and encourage the adoption of more virtuous behaviors.

Last but not least, international cooperation is crucial to dealing effectively with air pollution, which knows no borders. Michel Ramonet and Philippe Ciais (2021, p. 167) stress the importance of international agreements and scientific exchanges to harmonize air quality standards and share best practices between cities worldwide.

In conclusion, improving air quality and reducing air pollution in urban areas requires an integrated approach, combining regulatory measures, technological innovation, urban planning and citizen mobilization. These efforts must form part of a long-term strategy, taking into account the complex interactions between air quality, climate and public health.

### 5.1.4 Sustainable waste management and promotion of the circular economy

Sustainable waste management and the promotion of the circular economy have become crucial issues for 21st century cities, faced with growing waste production and the need to preserve natural resources.

This issue requires a systemic approach, integrating technical, economic, social and environmental aspects.

Dominique Bourg and Nicholas Buclet (2019, p. 56) highlight the importance of fundamentally rethinking our relationship with waste from a circular economy perspective. According to these authors, it is essential to move from a linear "extract-produce-consume-throw away" logic to a circular model where waste is considered as a resource. This transition implies an overhaul of production and consumption systems, as well as a change in individual and collective behavior.

Waste prevention and reduction at source are the first pillar of sustainable management. Jean-Michel Balet (2020, p. 123) analyzes the effectiveness of waste prevention policies implemented in several French cities. His study shows that measures such as promoting home composting, combating food waste and encouraging repair can significantly reduce the volume of waste produced. The author also stresses the importance of awareness-raising and education to encourage the adoption of eco-responsible behavior.

Optimizing waste collection and sorting is another crucial aspect of sustainable management. Hélène Beraud and Bruno Durand (2018, p. 89) highlight the value of new technologies in improving the efficiency of collection systems. In particular, their study analyzes the contribution of smart sensors and the Internet of Things to optimizing collection rounds and improving sorting rates. However, the authors emphasize the need to adapt these solutions to local contexts, and to take account of personal data protection issues.

The development of waste recycling and recovery is at the heart of the circular economy. Rémi Beulque and Franck Aggeri (2021, p. 178) explore the opportunities and challenges involved in setting up innovative recycling channels in urban environments. Their research highlights the importance of developing advanced sorting technologies, improving the quality of recycled materials and creating markets for these secondary materials. The authors stress the need for close collaboration between local authorities, manufacturers and players in the social economy to develop these channels.

Integrating the circular economy into urban planning is a major challenge for cities. Sabine Barles and Josefine Fokdal (2020, p. 234) analyze the concept of "urban metabolism" and its application to the management of material and energy flows on a city scale. Their study shows how a systems approach can identify synergies between different urban sectors and create closed resource loops, thereby reducing waste production and consumption of virgin resources.

Energy recovery from non-recyclable waste is an important component of sustainable management. Rémi Guillet and Isabelle Hebe (2017, p. 145) assess the environmental and economic performance of modern waste-to-energy plants. Their study shows that these facilities, when designed and operated according to the best available technologies, can make a significant contribution to the production of renewable energy while reducing greenhouse gas emissions associated with landfilling waste.

Social innovation and the collaborative economy offer new prospects for sustainable waste management. Valérie Guillard and Dominique Roux (2022, p. 67) explore the potential of citizen initiatives such as ressourceries, repair cafés and local exchange systems to extend the life of objects and reduce waste production. Their research highlights the importance of these initiatives in creating social links and raising awareness of the challenges of the circular economy.

Finally, the governance and financing of waste management are crucial aspects in ensuring the sustainability of systems. Mathieu Glachant and Nicolas Buclet (2019, p. 201) analyze the different incentive-based pricing models implemented in French cities. Their study shows that these schemes, when well designed and supported, can significantly reduce waste production and improve sorting performance, while ensuring a fair distribution of costs among users.

In conclusion, sustainable waste management and the promotion of the circular economy in urban environments require an integrated approach, combining technological innovation, behavioral change, appropriate urban planning and participatory governance. These efforts must be part of a long-term vision, aimed at transforming cities into sustainable and resilient urban ecosystems.

## 5.1.5 Protecting urban and peri-urban biodiversity

Protecting urban and peri-urban biodiversity has become a major challenge for 21st century cities, faced with the accelerating erosion of living organisms and the need to rethink their relationship with nature. This complex issue calls for an interdisciplinary approach, integrating ecological, urban planning, social and political aspects.

Philippe Clergeau and Nathalie Machon (2020, p. 45) highlight the importance of urban biodiversity for the functioning of ecosystems and the well-being of city dwellers. Their work highlights the role of urban green spaces, wastelands and micro-habitats in maintaining a diversity of flora and fauna in the city. The authors stress the need to consider the city as an ecosystem in its own right, with its own dynamics and complex interactions between species.

Urban ecological planning is an essential lever for protecting and restoring biodiversity. According to Luc Abbadie and Audrey Muratet (2019, p. 123), integrating the green and blue grid into urban planning documents makes it possible to preserve and reconnect natural habitats within the urban fabric. Their study shows that creating ecological corridors and preserving biodiversity reservoirs help maintain connectivity between species populations and foster their resilience in the face of urban pressures.

Differentiated management of urban green spaces plays a crucial role in protecting biodiversity. Gilles Lecuir and Grégoire Loïs (2018, p. 89) analyze the impact of these practices on the specific richness of urban environments. Their work shows that reducing the use of pesticides, diversifying plant strata and adapting maintenance periods to species' biological cycles can significantly increase biodiversity in urban parks and gardens.

Protecting endangered species in urban environments requires targeted action. Frédéric Jiguet and Romain Julliard (2021, p. 156) study the conservation measures implemented to preserve urban breeding bird populations. Their research highlights the effectiveness of measures such as the installation of suitable nesting boxes, the creation of quiet zones during the breeding season, and the ecological management of urban water bodies to encourage the nesting of endangered species.

The fight against invasive alien species is a major challenge for the protection of urban biodiversity. Emmanuelle Porcher and Anne-Caroline Prévot (2020, p. 234) analyze strategies for managing these species in urban environments. Their study underlines the importance of early detection and rapid intervention to limit their spread, while warning of the potentially negative effects of certain control methods on local ecosystems.

Involving citizens in the protection of urban biodiversity is essential. Vincent Pellissier and Cécile Clavel (2019, p. 178) explore the potential of participatory science to raise public awareness and collect data on urban biodiversity. Their work shows that these initiatives, such as programs to monitor garden birds or urban pollinators, contribute to both the acquisition of scientific knowledge and the appropriation of biodiversity issues by city dwellers.

Innovative building greening offers new prospects for urban biodiversity. Nathalie Blanc and Philippe Clergeau (2022, p. 67) study the impact of green roofs and facades on urban flora and fauna. Their research highlights the potential of these developments to create new habitats and encourage the presence of rare species in dense urban environments, while underlining the need for ecological design adapted to urban constraints.

Protecting peri-urban biodiversity is also crucial to maintaining connections between urban and rural ecosystems. Laurent Simon and Jean-Marc Berton (2017, p. 201) analyze the challenges of preserving agricultural and natural areas on the outskirts of cities. Their study highlights the importance of limiting urban sprawl, preserving green belts and promoting biodiversity-friendly peri-urban agriculture to maintain ecosystem services essential to cities.

Finally, adaptation to climate change must be integrated into strategies for protecting urban biodiversity. Sandra Lavorel and Thierry Tatoni (2021, p. 112) study the potential impacts of climate change on urban ecosystems and propose adaptation measures. Their work highlights the need to promote the genetic and functional diversity of urban species, create climate refuges and anticipate changes in species distribution to maintain the resilience of urban ecosystems.

In conclusion, protecting urban and peri-urban biodiversity requires an integrated approach, combining conservation, ecological restoration, urban planning and public awareness. These efforts must be part of a long-term vision, aimed at reconciling urban development with the preservation of living things, and creating more resilient and ecologically connected cities.

## 5.2 Energy efficiency, renewable energies and the green economy

### 5.2.1 Promoting energy efficiency in buildings and urban infrastructure

Promoting energy efficiency in buildings and urban infrastructure has become a crucial issue for 21st century cities, faced with the need to reduce their carbon footprint and optimize their energy consumption. This complex issue requires a multidimensional approach, integrating technical, economic, social and regulatory aspects.

Bruno Peuportier and Christophe Ménézo (2019, p. 56) highlight the importance of energy efficiency in the building sector, which accounts for a significant share of total energy consumption in cities. Their work highlights the considerable potential for energy savings associated with the thermal renovation of existing building stock and the construction of high-performance buildings. The authors stress the need to adopt a global approach, taking into account the entire life cycle of buildings, from design to demolition.

Energy renovation of existing housing stock is a major challenge for cities. Marie-Hélène Laurent and Dominique Osso (2020, p. 123) analyze the obstacles and levers for accelerating the energy renovation of private housing. Their study highlights the importance of financial incentives, technical support for homeowners and the structuring of competent professional sectors in overcoming the obstacles to renovation. The authors also highlight the key role played by local authorities in coordinating players and setting up support systems tailored to local contexts.

Technological innovation plays a crucial role in improving the energy efficiency of buildings. Etienne Wurtz and Laurent Mora (2018, p. 89) explore the potential of innovative materials and intelligent energy management systems to optimize the energy performance of buildings. Their research highlights the contribution of high-performance insulation, dynamic glazing and heat-recovery ventilation systems to reducing buildings' energy needs. However, the authors stress the need to assess the overall environmental impact of these technologies over their entire life cycle.

Integrating renewable energies into urban buildings is an important way of improving the overall energy efficiency of cities. Françoise Blanc and Philippe Bosseboeuf (2021, p. 178) analyze the opportunities and challenges associated with the development of building-integrated photovoltaics and district heating networks powered by renewable energies. Their study underlines the importance of energy planning at district or city level to optimize the production and distribution of renewable energy.

Bioclimatic building design offers interesting prospects for reducing energy requirements. Alain Guyot and Samuel Courgey (2020, p. 234) explore the principles of bioclimatic architecture and their application in the urban context. Their work shows how optimizing building orientation, thermal envelope design and the use of passive solar protection can significantly reduce heating and cooling requirements, while improving occupant comfort.

The energy efficiency of urban infrastructures is also a major issue. Sabine Barles and Jean-Pierre Lévy (2017, p. 145) analyze the energy optimization potential of urban networks (public lighting, water distribution, sanitation). Their study highlights the importance of modernizing equipment, optimizing processes and using smart technologies to reduce the energy consumption of urban infrastructures.

User awareness and support are essential to maximize the benefits of energy efficiency investments. Marie-Christine Zélem and Christophe Beslay (2019, p. 67) explore the social and behavioral dimensions of energy consumption in buildings. Their research highlights the importance of information, training and support for occupants to optimize equipment use and adopt virtuous energy behaviors.

Regulatory and economic tools play a crucial role in promoting energy efficiency. Bernard Laponche and Olivier Sidler (2022, p. 201) analyze the impact of thermal regulations and financial incentives on improving the energy performance of the building stock. Their study shows that the combination of ambitious standards and targeted incentives can significantly accelerate the transition to more energy-efficient buildings.

Finally, the district or urban block approach offers new perspectives for energy optimization. Charlotte Tardieu and Morgane Colombert (2020, p. 112) study the concept of thermal and electrical "smart grids" on a neighborhood scale. Their work highlights the potential of these systems to optimize the management of energy flows, promote collective self-consumption and reduce peak consumption on an urban scale.

In conclusion, promoting energy efficiency in buildings and urban infrastructure requires an integrated approach, combining technological innovation, regulatory changes, economic incentives and user support. These efforts must be part of a systemic vision of the city, taking into account the complex interactions between buildings, infrastructures and user behavior to create more sustainable and resilient urban environments.

## 5.2.2 Development of local renewable energies (solar, wind)

The development of renewable energies on a local scale, particularly solar and wind power, has become a crucial issue for cities in their transition to a more sustainable energy model. This complex issue requires a multidimensional approach, integrating technological, urban planning, economic and social aspects.

Daniel Lincot and Anne-Sophie Claeys-Mekdade (2019, p. 45) highlight the considerable potential of solar photovoltaics in urban environments. Their work highlights recent technological advances that have improved the efficiency and architectural integration of solar panels. The authors stress the need for strategic planning on a city-wide scale to optimize the exploitation of the solar deposit, taking into account constraints linked to shading and the preservation of built heritage.

Integrating solar power into existing buildings poses specific challenges. Marion Persem and Bruno Peuportier (2020, p. 123) analyze the technical and regulatory issues involved in installing photovoltaic systems on the roofs and facades of urban buildings. Their study highlights the importance of an integrated approach, taking into account structural, aesthetic and overall energy performance aspects of the building. The authors also highlight the potential of innovative technologies such as transparent photovoltaic cells and solar tiles to facilitate architectural integration.

The development of urban wind power presents specific opportunities and challenges. Jérôme Defossez and Emmanuel Rey (2018, p. 89) explore the potential of vertical-axis wind turbines and micro-wind turbines for energy production in dense urban environments. Their research highlights the advantages of these technologies in terms of compactness and adaptation to the turbulent wind regimes characteristic of urban environments. However, the authors stress the need for a precise assessment of local wind conditions and potential impacts on the neighborhood (noise, vibrations) to ensure the viability of projects.

Collective self-consumption and local energy communities open up new prospects for the development of renewable energies on a neighborhood scale. Marie Dégremont and Cécile Blatrix (2021, p. 178) analyze emerging models of collective organization for the production and consumption of renewable energy in cities. Their study highlights the potential of these initiatives to optimize the match between local production and consumption, while promoting citizen involvement in the energy transition.

Integrating renewable energies into existing urban grids poses technical and organizational challenges. Nouredine Hadjsaïd and Jean-Claude Sabonnadière (2020, p. 234) explore the issues involved in managing smart grids integrating a high proportion of intermittent renewable energies. Their work highlights the importance of storage technologies, active demand management and advanced forecasting systems in guaranteeing grid stability and reliability.

Territorial energy planning plays a crucial role in the coherent development of renewable energies on a local scale. Gilles Debizet and Stéphane La Branche (2019, p. 67) analyze urban energy planning tools and methods. Their research highlights the importance of a systemic approach, integrating analysis of renewable deposits, assessment of energy needs and consideration of urban planning and environmental constraints to define renewable energy development strategies tailored to local contexts.

Economic and financial aspects are crucial to the viability of urban renewable energy projects. Philippe Menanteau and Dominique Finon (2022, p. 201) examine the economic models and support mechanisms suited to the development of renewable energies in urban environments. Their analysis underlines the importance of targeted support schemes that take into account the specific features of urban projects (higher integration costs, space constraints) while encouraging innovation and cost optimization.

The social acceptability of renewable energy projects in urban environments is a major issue. Yves

Marignac and Sophia Majnoni d'Intignano (2020, p. 112) explore the factors influencing urban residents' perception of and support for local renewable energy projects. Their work highlights the importance of consultation, transparency and the involvement of local residents in the design and governance of projects, to promote ownership and long-term success.

Finally, technological innovation continues to open up new prospects for the integration of renewable energies in urban environments. Christophe Ballif and Nicolas Wyrsch (2021, p. 156) explore recent advances in new-generation solar cells (perovskites, tandem cells) and their potential to improve the efficiency and reduce the costs of urban solar production. Their research highlights the importance of continued R&D support to accelerate the deployment of these innovative technologies.

In conclusion, the development of local renewable energies, particularly solar and wind power, requires an integrated approach, combining technological innovation, strategic planning, adaptation of regulatory and economic frameworks, and citizen involvement. These efforts must be part of a global vision of urban energy transition, aimed at creating more energy-independent, resilient and sustainable cities.

## 5.2.3 Setting up sustainable urban transport systems and soft mobility

Implementing sustainable urban transport and

soft mobility systems has become a major challenge for 21st century cities, faced with the challenges of congestion, air pollution and the need to reduce greenhouse gas emissions. This complex issue requires a holistic approach, integrating urban planning, technological, social and political aspects.

Jean-Pierre Orfeuil and Caroline Gallez (2019, p. 56) highlight the importance of urban planning that promotes sustainable mobility. Their work highlights the close link between urban form and mobility practices, arguing for proximity-based urban planning that reduces travel distances and encourages active modes. The authors insist on the need to rethink urban planning to create more compact, mixed-use and multipolar cities, conducive to sustainable mobility.

The development of public transport is an essential pillar of sustainable urban transport systems. Aurélien Delpirou and Mathieu Flonneau (2020, p. 123) analyze strategies for strengthening and optimizing public transport networks in metropolises. Their study highlights the importance of intermodality, fare integration and improved service quality in making public transport more attractive than the private car. The authors also highlight the potential of Bus Rapid Transit (BRT) and modern tramways to effectively structure urban mobility.

Promoting active mobility, particularly cycling and walking, plays a crucial role in the transition to more sustainable transport systems. Frédéric Héran and

Emmanuel Ravalet (2018, p. 89) explore the levers for developing cycling in cities. Their research highlights the importance of a safe, continuous cycling network, a suitable parking offer and complementary services (self-service bikes, repair workshops) to encourage daily cycling. The authors also stress the need for a comprehensive policy in favor of active modes, including measures to moderate car traffic and redesign public spaces.

The electrification of urban transport offers interesting prospects for reducing pollutant and greenhouse gas emissions. Dominique Finon and Patrice Geoffron (2021, p. 178) analyze the issues involved in deploying electric vehicles and electrifying public transport fleets. Their study underlines the importance of an integrated approach, taking into account energy, infrastructure and economic aspects, to ensure the sustainability of this transition. The authors also warn of potential rebound effects and the need for decarbonized power generation to maximize environmental benefits.

New information and communication technologies (ICT) are playing an increasingly important role in optimizing urban transport systems. Gilles Pinson and Vincent Kaufmann (2020, p. 234) explore the potential of "smart mobility" to improve the efficiency and flexibility of urban travel. Their work highlights the contribution of mobile applications, real-time information systems and multimodal planning tools in facilitating travel and encouraging the use of alternatives to the private car.

Mobility demand management is an important lever for reducing travel requirements and optimizing the use of existing infrastructures. Marie- Hélène Massot and Jimmy Armoogum (2019, p. 67) analyze mobility management strategies such as telecommuting, flexible working hours and company travel plans. Their research highlights the importance of a coordinated approach between public authorities, employers and citizens to sustainably change mobility behavior.

Vehicle sharing and pooling offer interesting prospects for optimizing the use of resources and reducing the car's footprint in the city. Sylvie Fol and Caroline Gallez (2022, p. 201) study the development of car-sharing and urban car-pooling. Their analysis highlights the potential of these practices to reduce household motorization rates and optimize vehicle occupancy, while underlining the challenges of integrating these services into the overall urban mobility offer.

The governance of urban mobility is a crucial issue in ensuring the coherence and effectiveness of sustainable transport policies. Pierre-Henri Emangard and Bruno Faivre d'Arcier (2020, p. 112) explore innovative governance models, such as mobility organizing authorities (AOM) on a metropolitan scale. Their work highlights the importance of enhanced coordination between the various players (local authorities, transport operators, users) and an integrated approach to transport and urban planning.

Finally, the social acceptability and equity of sustainable mobility policies are essential aspects to be taken into account. Christophe Jemelin and Vincent Kaufmann (2021, p. 156) analyze the social issues involved in the transition to more sustainable transport systems. Their research highlights the need for inclusive policies, taking into account the specific needs of different social groups and territories, to guarantee equitable access to mobility while pursuing sustainability objectives.

In conclusion, the implementation of sustainable urban transport and soft mobility systems requires a systemic approach, combining actions on transport supply, urban planning, technologies and behaviours. These efforts must be part of a long-term vision of urban development, aimed at creating more liveable, accessible and environmentally-friendly cities.

### 5.2.4 Support for green economy initiatives and eco-industries

Support for green economy initiatives and eco-industries has become a major thrust of sustainable urban development policies. This approach aims to reconcile economic growth, innovation and environmental preservation, while creating local jobs and improving the quality of urban life. Implementing these policies raises complex issues, requiring a multi-dimensional approach.

Dominique Bourg and Christian Arnsperger (2019, p. 45) highlight the importance of the circular economy as a pillar of the urban green economy. Their work highlights the potential for value creation and reduced environmental impact linked to the establishment of recycling and resource reuse loops on a local scale. The authors stress the need for a systemic approach, integrating the entire value chain from product design to end-of-life.

The development of local eco-industries plays a crucial role in the ecological transition of cities. Patricia Crifo and Bénédicte Meurisse (2020, p. 123) analyze support strategies for innovative companies in the environmental and energy fields. Their study highlights the importance of targeted innovation policies, specialized incubators and thematic clusters in fostering the emergence and growth of local green businesses. The authors also highlight the key role of public-private partnerships in developing innovative solutions to urban environmental challenges.

The collaborative economy and citizen initiatives provide fertile ground for the emergence of green economy initiatives on a local scale. Valérie Guillard and Dominique Roux (2018, p. 89) explore the potential of sharing, repair and reuse practices to reduce resource consumption and create social ties in urban environments. Their research highlights the importance of local government support for these citizen initiatives, through the provision of spaces, legal support and the promotion of collaborative practices.

The energy transition offers major opportunities for the development of local eco-industries. Philippe Quirion and Céline Guivarch (2021, p. 178) analyze the potential for job creation linked to the development of renewable energies and energy-efficient building renovation in urban areas. Their study underlines the importance of appropriate training and retraining policies to meet the skills needs of emerging green industries.

Urban and peri-urban agriculture is a promising area for the local green economy. Christine Aubry and Jeanne Pourias (2020, p. 234) explore innovative economic models linked to urban food production, from rooftop agriculture to vertical farms. Their work highlights the potential of these initiatives to create local jobs, improve urban food security and reduce the carbon footprint associated with supplying food to cities.

Sustainable waste management offers interesting prospects for the development of local eco-industries. Jean-Marc Meunier and Sabine Barles (2019, p. 67) analyze emerging urban waste recovery processes, from composting and upcycling to methanization. Their research highlights the importance of an integrated approach to waste management, combining reduction at source, reuse and recovery, to maximize environmental and economic benefits.

The development of eco-construction and bio-sourced materials represents a promising avenue for urban eco-industries. Bruno Peuportier and Nadia Hoyet (2022, p. 201) examine innovations in green building materials and sustainable construction techniques. Their analysis highlights the potential of these innovative approaches to create local jobs and reduce the carbon footprint of the building sector.

Sustainable mobility also offers significant opportunities for the development of local eco-industries. Frédéric Héran and Francis Papon (2020, p. 112) explore emerging industries linked to electric vehicles, bicycles and new shared mobility services. Their work highlights the importance of a local ecosystem of innovation and production to maximize the economic benefits of the transition to more sustainable mobility.

Last but not least, the development of green technologies and digital solutions for sustainable cities is a major focus of the urban green economy. Gilles Pinson and Antoine Picon (2021, p. 156) analyze the potential of smart cities to optimize resource management and reduce the environmental impact of cities. Their research highlights the importance of an ethical and inclusive approach to technological development, taking into account issues of data protection and accessibility for all.

In conclusion, supporting green economy initiatives and eco-industries requires an integrated approach, combining innovation policies, support for entrepreneurship, vocational training and civic involvement. These efforts must be part of an overall vision of sustainable urban development, aimed at creating resilient, innovative and environmentally-friendly local economic ecosystems. The success of these policies relies on close collaboration between local authorities, the private sector, research and civil society to co-construct solutions tailored to the specific challenges of each urban area.

## 5.2.5 Raising awareness and educating citizens about the environment

Citizen awareness-raising and environmental education are crucial to the success of sustainable urban development policies. These initiatives aim to inform, empower and involve residents in their city's ecological transition. Implementing effective strategies in this area raises complex issues, requiring a multi-dimensional approach tailored to different audiences.

Lucie Sauvé and Isabel Orellana (2019, p. 45) emphasize the importance of a holistic approach to environmental education in urban settings. Their work highlights the need to go beyond the simple transmission of information to develop genuine, active "eco-citizenship". The authors emphasize the importance of experiential learning and local anchoring of educational approaches to encourage real appropriation of environmental issues by city dwellers.

Environmental communication plays a key role in raising public awareness. Thierry Libaert and Jean-Marie Pierlot (2020, p. 123) analyze effective communication strategies for promoting eco-responsible behavior in the city. Their study highlights the importance of positive communication, focusing on the individual and collective benefits of actions in favor of the environment, rather than on guilt-tripping rhetoric. The authors also highlight the potential of new technologies and social networks to reach a wide audience and encourage civic engagement.

Involving citizens in concrete sustainable development projects is a powerful lever for awareness-raising and education. Hélène Subrémon and Gaëtan Brisepierre (2018, p. 89) explore the potential of participatory approaches to the urban environment. Their research highlights the importance of learning by doing and the sense of empowerment generated by these initiatives in fostering lasting behavioral change.

Environmental education in schools plays a fundamental role in training future eco-responsible citizens. Jean-Marc Lange and Yves Girault (2021, p. 178) analyze innovative pedagogical approaches to integrating sustainable development issues into school curricula. Their study underlines the importance of a cross-disciplinary approach, linking environmental issues to other disciplines, and of an active pedagogy that encourages critical thinking and student involvement.

Participatory science initiatives offer interesting opportunities for raising citizens' awareness of urban environmental issues. Françoise Gourmelon and

Matthieu Noucher (2020, p. 234) explore the potential of citizen-based environmental data collection projects to develop an active ecological conscience. Their work highlights the importance of these approaches in creating a direct link between residents and their environment, while contributing to the production of scientific knowledge useful for urban environmental management.

Training adults in environmental issues is also crucial to support the ecological transition of cities. Emmanuel Rivat and Cécile Renouard (2019, p. 67) analyze continuing education and retraining schemes in the environmental and sustainable development professions. Their research highlights the importance of a diversified and accessible training offer to enable citizens to adapt to changes in the job market linked to the ecological transition.

Digital tools offer new perspectives for environmental awareness and education. Éric Avenel and Olivier Petit (2022, p. 201) study the potential of mobile applications, serious games and online platforms to promote eco-responsible behavior in urban environments. Their analysis highlights the value of these tools in personalizing messages, fostering long-term commitment and creating communities of active eco-citizens.

Art and culture are powerful vectors for raising awareness of environmental issues. Nathalie Blanc and Julie Celnik (2020, p. 112) explore the role of artistic interventions and cultural events in raising ecological awareness in urban settings. Their work underlines the importance of these sensitive approaches in reaching a wide audience and sparking critical reflection on our relationship with the environment.

Finally, evaluating and monitoring the impact of environmental education and awareness campaigns is essential to improving their effectiveness. Cécile Blatrix and Jacques Méry (2021, p. 156) analyze evaluation methodologies for urban environmental education programs. Their research highlights the importance of an approach combining quantitative and qualitative indicators to measure not only changes in knowledge, but also actual changes in behavior and impact on the quality of the urban environment.

In conclusion, raising citizens' environmental awareness and education in urban areas requires a comprehensive approach, combining information, experimentation, participation and concrete commitment. These efforts must be part of a long-term strategy, adapted to local conditions and different audiences, to create a genuine culture of urban sustainability. The success of these approaches depends on close collaboration between local authorities, the educational world, associations, the media and citizens themselves, to co-construct a shared vision of the sustainable city and the means to achieve it.

## 5.3 Resilient urban planning and land use

### 5.3.1 Integrating climate issues into urban planning documents

The integration of climate issues into urban planning documents has become an imperative for cities in the face of the challenges posed by climate change. The aim is to anticipate and mitigate the impacts of global warming, while adapting the urban fabric to new climatic conditions. Implementing this integration raises complex issues, requiring a multidisciplinary approach and a thorough review of urban planning practices.

Vincent Béal and Max Rousseau (2019, p. 45) highlight the importance of a systemic approach to integrating climate issues into urban planning documents. Their work highlights the need to move beyond sectoral approaches to adopt a global vision, taking into account the interactions between climate, urban planning, mobility, energy and biodiversity. The authors stress the importance of multi-level governance to ensure coherence between the different planning scales, from the neighborhood to the urban region.

Taking climate scenarios into account in urban planning is a crucial issue. Valéry Masson and Aude Lemonsu (2020, p. 123) analyze the methodologies used to integrate regionalized climate projections into urban planning documents. Their study highlights the importance of close collaboration between climatologists and urban planners to translate scientific data into concrete planning guidelines. The authors also emphasize the need for a dynamic approach, enabling planning documents to be regularly adjusted in line with evolving climatic knowledge.

Combating urban heat islands is a major focus for integrating climate issues into urban planning. Jean-Pierre Lévy and Samuel Royer (2018, p. 89) explore planning strategies for reducing the heat island effect through urban planning documents. Their research highlights the importance of prescriptions relating to revegetation, water management and building materials in creating more favorable urban microclimates. The authors also highlight the key role played by urban morphology in the thermal regulation of cities.

The management of climatic risks, particularly floods and droughts, must be fully integrated into urban planning documents. Magali Reghezza-Zitt and Samuel Rufat (2021, p. 178) analyze approaches to anticipating and managing these risks through urban planning. Their study stresses the importance of accurate mapping of at-risk zones and the adoption of appropriate urban planning rules, such as limiting soil sealing or establishing flood expansion zones.

Integrating energy and climate issues into urban planning documents is essential to reducing greenhouse gas emissions. Olivier Coutard and Jonathan Rutherford (2020, p. 234) explore strategies for promoting energy efficiency and the development of renewable energies through urban planning. Their work highlights the importance of prescriptions relating to the energy performance of buildings, the orientation of construction and the establishment of district heating networks.

Sustainable mobility is a crucial aspect of integrating climate issues into urban planning. Caroline Gallez and Hélène Reigner (2019, p. 67) analyze the levers for reducing car dependency and promoting low-carbon modes of travel through urban planning documents. Their research highlights the importance of integrated urban and transport planning, favoring densification around public transport routes and the creation of mixed-use, compact neighborhoods.

Preserving and strengthening the urban green and blue fabric plays an essential role in adapting to climate change. Philippe Clergeau and Nathalie Blanc (2022, p. 201) examine approaches to integrating biodiversity and ecosystem service issues into urban planning documents. Their analysis highlights the importance of prescriptions relating to green spaces, stormwater management and ecological continuity in strengthening urban resilience to climate change.

The link between urban planning documents and territorial climate-air-energy plans (PCAET) is a major challenge in ensuring the coherence of local policies. François Bertrand and Elsa Richard (2020, p. 112) explore how the climate objectives of PCAETs can be integrated into urban planning documents. Their work highlights the importance of a cross-functional approach, enabling emission reduction and climate change adaptation objectives to be applied to all components of urban planning.

Lastly, citizen participation in the development of urban planning documents integrating climate issues is essential to ensure their appropriation and effective implementation. Hélène Reigner and Séverine Frère (2021, p. 156) analyze the participatory approaches used to involve residents in the definition of climate orientations in urban planning. Their research highlights the importance of these approaches in raising citizens' awareness of climate issues and encouraging the emergence of innovative solutions adapted to local contexts.

In conclusion, integrating climate issues into urban planning documents requires a comprehensive, cross-disciplinary approach, mobilizing a wide range of expertise and involving all stakeholders. This approach implies a thorough overhaul of urban planning practices, placing climate resilience at the heart of urban planning choices. The success of this integration depends on a long-term vision, adaptive governance and the mobilization of all urban planning levers to create cities that are more sustainable and resilient in the face of climate change.

## 5.3.2 Development of ecological neighborhoods and compact cities

The development of ecological neighborhoods and compact cities has become a major focus of sustainable urban policies. This approach aims to reconcile urban density, quality of life and environmental performance, while reducing urban sprawl and its negative impacts. Implementing these concepts raises complex issues, requiring an integrated and innovative approach to urban planning.

Cyria Emelianoff and Taoufik Souami (2019, p. 45) emphasize the importance of a holistic approach to the design of ecological neighborhoods. Their work highlights the need to go beyond purely technical approaches to integrate the social, economic and cultural dimensions of sustainable development. The authors stress the importance of involving local residents in the design and management of these neighborhoods to ensure their appropriation and sustainability.

Urban densification, a pillar of the compact city, raises major challenges in terms of social acceptability and urban quality. Vincent Fouchier and Jean-Michel Roux (2020, p. 123) analyze strategies for reconciling density and quality of life in urban projects. Their study highlights the importance of a qualitative approach to densification, focusing on functional mix, quality of public spaces and the presence of nature in the city to create attractive, liveable urban environments.

The integration of energy issues is crucial to the development of green neighborhoods. Olivier Coutard and Jonathan Rutherford (2018, p. 89) explore innovative approaches to energy production and management at the neighborhood scale. Their research highlights the potential of smart grids, district heating networks and collective self-consumption to optimize energy efficiency and reduce the carbon footprint of neighborhoods.

Sustainable water and waste management is a major focus of ecological neighborhoods. Bernard Barraqué and Rémi Barbier (2021, p. 178) analyze solutions for reducing water consumption, promoting recycling and managing rainwater in an environmentally-friendly way. Their study underlines the importance of an integrated approach, combining innovative technologies and changes in user behavior.

Sustainable mobility is at the heart of the compact city concept. Caroline Gallez and Hélène Reigner (2020, p. 234) explore strategies for reducing car dependency and promoting active and collective modes of travel in ecological neighborhoods. Their work highlights the importance of integrated urban and transport planning, favoring proximity to services and multimodal accessibility.

Urban biodiversity plays an essential role in the environmental quality of green neighborhoods. Philippe Clergeau and Nathalie Blanc (2019, p. 67) analyze approaches to integrating nature in the city and enhancing ecosystem services in urban projects. Their research highlights the importance of a systemic approach to the urban green and blue grid, combining green spaces, green roofs and ecological management of public spaces.

Bioclimatic architecture and eco-construction are key components of ecological neighborhoods. Bruno Peuportier and Nadia Hoyet (2022, p. 201) examine innovations in architectural design and sustainable materials to reduce the environmental impact of buildings. Their analysis highlights the importance of a global approach, integrating energy performance, user comfort and the life cycle of materials.

Social and functional diversity are crucial to the sustainability of ecological neighborhoods and compact cities. Marie-Hélène Bacqué and Sylvie Fol (2020, p. 112) explore strategies for promoting the diversity of populations and activities in sustainable urban projects. Their work underlines the importance of proactive policies in terms of social housing, public facilities and urban animation in creating lively, inclusive neighborhoods.

The circular economy and short circuits are promising avenues for strengthening the sustainability of eco-districts. Sabine Barles and Josefine Fokdal (2021, p. 156) analyze approaches to developing local, resilient economic ecosystems on a neighborhood scale. Their research highlights the importance of collaborative economy initiatives, local production spaces and proximity exchange systems in reducing the urban ecological footprint.

Finally, the governance and financing of eco-neighborhood and compact city projects raise specific issues. Gilles Novarina and Natacha Seigneuret (2019,

p. 289) explore innovative models of public-private partnership and citizen participation in the design and management of these sustainable urban projects. Their work highlights the importance of a collaborative and adaptive approach, enabling projects to be adjusted to changing local needs and constraints.

In conclusion, the development of ecological neighborhoods and compact cities requires an integrated approach, combining technological, social and organizational innovations. These sustainable urban projects must be part of a long-term vision of territorial development, articulating the scales of the building, the neighborhood and the conurbation. The success of these approaches hinges on close collaboration between public, private and civic players, as well as on the ability to adapt solutions to local specificities while drawing inspiration from international best practices. The major challenge is to generalize these innovative approaches to the entire urban fabric, to create more sustainable, resilient and livable cities.

### 5.3.3 Adapting urban infrastructures to climate change

Adapting urban infrastructures to climate change has become a crucial issue in ensuring the resilience and sustainability of cities in the face of growing environmental challenges. This involves a thorough review of urban infrastructure design, construction and management practices to anticipate and mitigate the impacts of global warming. Implementing this adaptation raises complex issues, requiring a multidisciplinary and innovative approach.

Sébastien Blandin and Jean-Pierre Mignot (2019, p. 45) highlight the importance of a systems approach to urban infrastructure adaptation. Their work highlights the need to consider the interdependencies between different networks (water, energy, transport, telecommunications) to develop coherent and effective adaptation strategies. The authors stress the importance of integrated governance to ensure coordination between the various players involved in infrastructure management.

Managing the risks associated with extreme climatic events is a major focus of urban infrastructure adaptation. Magali Reghezza-Zitt and Samuel Rufat (2020, p. 123) analyze approaches for strengthening the resilience of urban networks to floods, heatwaves and storms. Their study highlights the importance of detailed vulnerability mapping and the adoption of flexible, redundant solutions to ensure the continuity of essential services in the event of a crisis.

Adapting water management infrastructures is crucial in the face of climate change. Bernard Chocat and Jean-Claude Deutsch (2018, p. 89) explore innovative strategies for dealing with the challenges of resource scarcity and intensifying rainfall events. Their research highlights the importance of integrated stormwater management approaches, combining technical solutions (retention basins, landscaped valleys) and resilient urban planning (soil desoiling, revegetation).

Transport infrastructures are particularly vulnerable to the impacts of climate change. Yves Crozet and Alain Bonnafous (2021, p. 178) analyze the challenges of adapting road, rail and airport networks to new climatic conditions. Their study underlines the importance of an anticipatory approach, incorporating climate projections into infrastructure design and maintenance, as well as the development of more flexible and resilient transport systems.

The adaptation of energy networks is essential to ensure security of supply in a context of climate change. Olivier Coutard and Jonathan Rutherford (2020, p. 234) explore strategies for enhancing the resilience of energy production, distribution and consumption systems. Their work highlights the importance of smart grids, diversifying energy sources and adapting infrastructures to new climatic conditions (heat resistance, flood protection).

Green infrastructure is playing a growing role in urban adaptation to climate change. Philippe Clergeau and Nathalie Blanc (2019, p. 67) analyze the potential of nature-based solutions to strengthen urban resilience. Their research highlights the importance of green and blue grids, green roofs and façades, and open spaces to regulate the urban microclimate, manage stormwater and preserve biodiversity.

Adapting buildings and public spaces to new climatic conditions is a major challenge for the comfort and health of city dwellers. Jean-Jacques Terrin and Jean-Baptiste Marie (2022, p. 201) examine innovative approaches to bioclimatic design, adapted materials and dynamic management of urban spaces. Their analysis highlights the importance of an integrated approach, combining passive and active solutions to create resilient and comfortable urban environments.

Digital infrastructures are playing a growing role in urban adaptation to climate change. Emmanuel Eveno and Gabriel Dupuy (2020, p. 112) explore the potential of information and communication technologies to improve risk management, optimize resources and strengthen urban resilience. Their work highlights the importance of early warning systems, modeling and simulation tools, and integrated infrastructure management platforms.

The adaptation of waste management infrastructures is also crucial in the face of climate challenges. Sabine Barles and Josefine Fokdal (2021, p. 156) analyze approaches to reducing the environmental impact of waste management and developing the circular economy on an urban scale. Their research highlights the importance of waste sorting, recycling and energy recovery infrastructures in reducing greenhouse gas emissions and optimizing the use of resources.

Finally, financing the adaptation of urban

infrastructure raises major issues. Nicolas Stern and Charlotte Taylor (2019, p. 289) explore innovative mechanisms for financing adaptation, combining public investment, public-private partnerships and green financial instruments. Their work highlights the importance of a long-term approach, integrating the costs avoided through adaptation into cost-benefit analyses of infrastructure projects.

In conclusion, adapting urban infrastructures to climate change requires a comprehensive, cross-disciplinary approach, mobilizing a wide range of expertise and involving all stakeholders in the area. This implies an in-depth review of infrastructure planning, design and management practices, placing climate resilience at the heart of investment choices. The success of this adaptation depends on a long-term vision, adaptive governance and the ability to innovate in order to develop flexible, sustainable solutions. The major challenge is to transform existing infrastructures while anticipating future needs, to create more resilient cities capable of adapting to future climate change.

## 5.3.4 Managing urban heat islands and greening public spaces

Managing urban heat islands and greening public spaces have become major priorities in strategies to adapt cities to climate change. These approaches aim to improve thermal comfort, quality of life and urban resilience in the face of increasingly frequent and intense heat waves. Implementing these strategies raises complex issues, requiring an integrated, multidisciplinary approach.

Valéry Masson and Aude Lemonsu (2019, p. 45) emphasize the importance of a detailed understanding of the mechanisms behind the formation of urban heat islands in order to develop effective mitigation strategies. Their work highlights the impact of urban morphology, surface materials and anthropogenic activities on the urban microclimate. The authors stress the importance of a systemic approach, integrating the interactions between buildings, vegetation, water and atmosphere into the design of urban spaces.

The greening of public spaces plays a crucial role in urban thermal regulation. Philippe Clergeau and Nathalie Blanc (2020, p. 123) analyze the multiple benefits of nature in the city in mitigating heat islands. Their study highlights the importance of diverse forms of vegetation (street trees, parks, pocket gardens, green roofs and facades) in maximizing ecosystem services and creating cool corridors throughout the city.

Urban water management is closely linked to the issue of heat islands. Bernard Chocat and Jean-Claude Deutsch (2018, p. 89) explore innovative approaches to integrated stormwater management for cooling the city. Their research highlights the potential of nature-based solutions, such as landscape valleys, rain gardens and infiltration basins, to combine stormwater management and the creation of cool islands.

Adapting buildings and public spaces to extreme

heat is a major challenge for bioclimatic urban planning. Jean-Jacques Terrin and Jean-Baptiste Marie (2021, p. 178) analyze design strategies for improving indoor and outdoor thermal comfort. Their study highlights the importance of high solar reflectivity materials, shading devices, natural ventilation and thermal inertia in creating cooler, more comfortable urban environments.

Urban planning plays a key role in the long-term management of heat islands. Vincent Béal and Max Rousseau (2020, p. 234) explore approaches to integrating climate issues into urban planning documents. Their work highlights the importance of prescriptions relating to revegetation, water management and urban form in creating cities that are more resilient to heat waves.

Green infrastructure on a city scale is a major lever for mitigating heat islands. Emmanuel Boutefeu and Antoine Brès (2019, p. 67) analyze the potential of urban green and blue grids to create interconnected cool networks. Their research highlights the importance of a multifunctional approach, combining biodiversity, stormwater management and the creation of recreational spaces to maximize the benefits of urban greening.

Citizen participation in the greening of public spaces is essential to ensure the success and sustainability of these initiatives. Hélène Reigner and Séverine Frère (2022, p. 201) examine participatory approaches to involving residents in the creation and maintenance of urban green spaces. Their analysis highlights the importance of these approaches in raising citizens' awareness of climate issues and encouraging them to take ownership of green spaces.

Urban agriculture is emerging as an innovative solution for combining greening, local food production and heat island mitigation. Christine Aubry and Jeanne Pourias (2020, p. 112) explore the potential of shared gardens, urban farms and productive roofs to create multifunctional cool islands. Their work highlights the importance of an integrated approach, combining food production, social ties and ecosystem services.

Managing urban green spaces in a context of climate change raises specific challenges. Marianne Cohen and Nathalie Frascaria-Lacoste (2021, p. 156) analyze strategies for adapting the plant palette and management practices to new climatic conditions. Their research highlights the importance of an anticipatory approach, favoring species adapted to hydric and thermal stress, and promoting diversity to strengthen the resilience of urban ecosystems.

Finally, assessing and monitoring the effects of revegetation on the urban microclimate are essential for optimizing adaptation strategies. Julien Bigorgne and Morgane Colombert (2019, p. 289) explore methodologies and tools for measuring and modeling the impact of revegetation interventions on heat islands. Their work highlights the importance of a rigorous scientific approach to guide public policy and assess the effectiveness of implemented measures.

In conclusion, managing urban heat islands and greening public spaces requires an integrated approach, combining scientific expertise, technical innovation and civic commitment. These strategies must be part of a long-term vision of urban development, articulating the scales of the building, the neighborhood and the conurbation. The success of these approaches depends on close collaboration between urban planners, ecologists, climatologists and city managers, as well as on the ability to adapt solutions to local conditions while drawing on international best practices. The major challenge is to generalize these approaches to the entire urban fabric, to create cities that are cooler, greener and more resilient in the face of climate change.

## 5.3.5 Promoting sustainable urban and peri-urban agriculture

The promotion of sustainable urban and peri-urban agriculture has become a major issue in the context of the ecological transition of cities and the search for more resilient food systems. This approach aims to bring production and consumption closer together, strengthen local food security, and generate multiple environmental and social benefits. Implementing these initiatives raises complex challenges, requiring an integrated and innovative approach.

Christine Aubry and Jeanne Pourias (2019, p. 45)

emphasize the importance of a systemic vision of urban and peri-urban agriculture. Their work highlights the diversity of forms this agriculture can take, from shared gardens to vertical farms, productive rooftops and peri-urban micro-farms. The authors stress the need to consider these different forms as complementary, each responding to specific objectives and contexts.

Integrating agriculture into the urban fabric raises important planning issues. Ségolène Darly and Monique Poulot (2020, p. 123) analyze strategies for preserving and enhancing agricultural spaces in urban and peri-urban environments. Their study highlights the importance of urban planning tools, such as protected agricultural zones and agri-urban projects, in ensuring the sustainability of these areas in the face of land pressure.

The sustainability of agricultural practices in urban and peri-urban environments is a crucial issue. Nicolas Bricas and Damien Conaré (2018, p. 89) explore agroecological approaches adapted to the urban context. Their research highlights the importance of low-input practices, the valorization of urban organic waste, and the preservation of cultivated biodiversity in developing resilient production systems with low environmental impact.

Access to land is a major challenge for the development of urban and peri-urban agriculture. Terre de Liens and SAFER (2021, p. 178) analyze innovative ways of facilitating the installation of farmers in urban and peri-urban areas. Their study underlines the importance of public land management tools, environmental rural leases and cooperative forms of agricultural land management in fostering the emergence of sustainable agricultural projects.

The social and educational dimension of urban agriculture is an essential aspect of its promotion. Flaminia Paddeu and Antoine Lagneau (2020, p. 234) explore the potential of shared gardens and educational farms to strengthen social ties, environmental education and awareness of food issues. Their work highlights the importance of these spaces as places for experimenting with new forms of citizenship and urban solidarity.

The integration of agriculture into architecture and urban planning is opening up new perspectives. Vertical Farm Institute and SOA Architects (2019, p. 67) analyze the potential of vertical farms, rooftop greenhouses and productive facades to intensify food production in dense urban environments. Their research highlights the importance of an integrated approach, combining agricultural production, energy management and the circular economy to maximize the benefits of these innovative systems.

Enhancing the value of short distribution channels and the social and solidarity economy is a major focus for the development of sustainable urban and peri-urban agriculture. Yuna Chiffoleau and Dominique Paturel (2022, p. 201) study innovative economic models that strengthen links between urban producers and local consumers. Their analysis highlights the importance of AMAPs, cooperative grocery stores and digital direct sales platforms in creating resilient and equitable local food systems.

Managing the health and environmental risks associated with urban agriculture requires special attention. Laurence Huc and Nathalie Gourmelen (2020, p. 112) explore the issues of soil pollution, air quality and the health safety of products from urban agriculture. Their work highlights the importance of a preventive approach, combining site diagnosis, choice of suitable crops and good agronomic practices to guarantee the quality and safety of urban produce.

The link between urban and peri-urban agriculture and territorial food policies is essential. Caroline Brand and Nicolas Bricas (2021, p. 156) analyze strategies for integrating urban and peri-urban agriculture into territorial food projects. Their research highlights the importance of local food governance, bringing together public, private and citizen players to develop sustainable food systems on a territorial scale.

Finally, evaluating and monitoring the impacts of urban and peri-urban agriculture is crucial to guiding public policy. François Mancebo and Sylvie Salles (2019, p. 289) explore methodologies and indicators for measuring the environmental, social and economic benefits of these initiatives. Their work highlights the importance of a multi-criteria approach, integrating the dimensions of food security, biodiversity, climate regulation and social well-being to assess the overall sustainability of urban and peri-urban agriculture.

In conclusion, promoting sustainable urban and peri-urban agriculture requires an integrated approach, mobilizing a wide range of expertise and involving all local players. This implies a review of urban planning, land management and food policy practices to create an environment conducive to the development of these initiatives. The success of this promotion depends on a long-term vision, participative governance and the ability to innovate in order to develop agricultural models adapted to the urban context. The major challenge is to move from one-off initiatives to a genuine territorial strategy, making urban and peri-urban agriculture a pillar of cities' ecological and food transition.

## 5.4 Natural disaster risk reduction and warning systems

### 5.4.1 Mapping risk zones and regulating land use

Mapping risk zones and regulating land use are essential elements in the management of natural and technological risks in urban areas. These tools make it possible to anticipate potential hazards, limit the exposure of people and property, and steer land-use planning towards greater resilience. Implementing these approaches raises complex issues, requiring close collaboration between scientific experts, political decision-makers and local players.

Magali Reghezza-Zitt and Samuel Rufat (2019, p.

45) emphasize the importance of a multidimensional approach to risk mapping. Their work highlights the need to consider not only the hazard (potentially dangerous natural or technological phenomenon), but also the vulnerability of exposed issues and the territory's resilience capacities. The authors stress the importance of dynamic mapping, taking into account the evolution of risks in the context of climate change and territorial mutations.

Regulating land use according to identified risks is a major prevention tool. Bruno Ledoux and Gilles Hubert (2020, p. 123) analyze the various regulatory instruments used to control urbanization in high-risk areas. Their study highlights the importance of the Plans de Prévention des Risques (PPR) in France, which define specific zoning and prescriptions for development and construction in risk-prone areas.

Integrating risks into urban planning documents is crucial for effective preventive management. Hélène Celdran and Vincent Beccaria (2018, p. 89) explore the modalities of articulation between risk mapping and urban planning. Their research highlights the importance of a transversal approach, integrating risk issues into Schémas de Cohérence Territoriale (SCoT) and Plans Locaux d'Urbanisme (PLU) to steer urban development towards less exposed areas.

Taking account of uncertainties in risk mapping is a major challenge. Eric Leroi and Frédéric Leone (2021, p. 178) analyze probabilistic approaches and multi-risk scenarios to better understand the complexity and variability of hazardous phenomena. Their study underlines the importance of transparent communication on uncertainties to inform public decision-making and raise public awareness of potential risks.

Adapting regulations to local conditions is essential for effective risk management. Freddy Vinet and Stéphanie Defossez (2020, p. 234) explore the methods of consultation and negotiation used to adjust regulatory prescriptions to local realities. Their work highlights the importance of a participatory approach, bringing together local elected representatives, experts and citizens to draw up land-use rules that are appropriate and accepted by all.

The management of already urbanized areas at risk raises particular issues. Anne-Laure Moreau and Jean- Michel Tanguy (2019, p. 67) analyze strategies for reducing the vulnerability of existing buildings and adapting uses in exposed areas. Their research highlights the importance of mitigation measures, such as reinforcing structures, adapting networks and setting up warning systems, in reducing potential damage.

The link between risk mapping and urban renewal policies is a major challenge for cities. Yves Raibaud and Hélène Reigner (2022, p. 201) examine approaches to reconciling urban densification and risk prevention. Their analysis highlights the importance of thinking in terms of urban form, favoring resilient and adaptable developments in moderately exposed areas.

Taking account of emerging risks and domino effects in mapping and regulations is essential. Valérie November and Marion Lecourt (2020, p. 112) explore methodologies for anticipating new risks linked to climate change, technological developments and interdependencies between urban systems. Their work underlines the importance of a forward-looking, systemic approach to the ongoing adaptation of risk management tools.

Economic risk assessment and cost-benefit analysis of prevention measures are crucial for guiding public policy. Nicolas Treich and Stéphane Hallegatte (2021, p. 156) analyze methodologies for quantifying the potential costs of disasters and the benefits of investment in prevention. Their research highlights the importance of a long-term approach, incorporating the costs avoided through land-use regulation into economic analyses.

Last but not least, it is essential that local players take ownership of the regulations and apply them effectively. Julie Daluzeau and Séverine Durand (2019, p. 289) explore the factors favoring or hindering the implementation of risk-related prescriptions. Their work highlights the importance of awareness-raising, training and support for local players to ensure the effectiveness of preventive measures.

In conclusion, the mapping of risk zones and the regulation of land use require an integrated approach, combining scientific expertise, legal tools and local consultation. These approaches must be part of a long-term vision of land use planning, combining risk prevention and sustainable development. The success of these approaches depends on close collaboration between government departments, local authorities, experts and citizens, as well as on the ability to continuously adapt tools to evolving knowledge and local contexts. The major challenge is to strike a balance between population protection and territorial development, to create cities that are more resilient in the face of natural and technological hazards.

## 5.4.2 Strengthening critical infrastructures to cope with climatic hazards

Reinforcing critical infrastructures in the face of climatic hazards has become a major challenge to ensure the resilience of cities and regions in a context of climate change. These infrastructures, which are essential to the functioning of society (energy, water, transport and telecommunications networks, etc.), are particularly vulnerable to extreme weather events, the frequency and intensity of which are increasing. Implementing reinforcement strategies raises complex technical, economic and organizational challenges.

Michel Lussault and Magali Reghezza-Zitt (2019, p. 45) emphasize the importance of a systems approach when analyzing the vulnerability of critical infrastructures. Their work highlights the interdependencies between different networks and the need to consider potential cascading effects in the event of a failure. The authors stress the importance of a holistic vision of urban resilience, integrating technical, social and organizational dimensions.

Adapting infrastructure design norms and standards is a key element of reinforcement. Jean-Marc Tacnet and Didier Richard (2020, p. 123) analyze approaches to integrating climate projections into technical standards. Their study highlights the importance of regularly revising standards to take account of evolving knowledge of climate change and its impacts.

System redundancy and diversification are key strategies for increasing resilience. Bruno Barroca and Damien Serre (2018, p. 89) explore how to design multi-scale, multi-source networks capable of maintaining minimal service in the event of disruption. Their research highlights the importance of decentralized solutions and backup systems in reducing the overall vulnerability of infrastructures.

Integrating nature-based solutions into infrastructure reinforcement offers promising prospects. Nathalie Blanc and Philippe Clergeau (2021, p. 178) analyze the potential of green and blue infrastructure to reduce the risks associated with flooding, heatwaves and erosion. Their study underlines the importance of a multifunctional approach, combining risk management and improvement of urban environmental quality.

Adaptive infrastructure management is crucial in the face of climate uncertainty. Guillaume Simonet and Laurent Bopp (2020, p. 234) explore flexible management approaches that enable reinforcement strategies to be adjusted in line with changing climatic conditions and scientific knowledge. Their work highlights the importance of monitoring and early warning systems for anticipating and responding to climate impacts.

Financing the reinforcement of critical infrastructures is a major challenge. Nicolas Stern and Joseph Stiglitz (2019, p. 67) analyze innovative mechanisms for financing adaptation, such as green bonds and public-private partnerships adapted to climate challenges. Their research highlights the importance of an economic assessment of climate risks to justify investments in resilience.

Coordination between different infrastructure managers is essential for a coherent approach to reinforcement. Céline Lutoff and Isabelle Ruin (2022, p. 201) examine the governance arrangements that can improve cooperation between public and private players in the management of critical infrastructures. Their analysis highlights the importance of information-sharing platforms and simulation exercises for strengthening coordination in crisis situations.

The integration of cybersecurity issues into the reinforcement of critical infrastructures has become essential. Eric Rigaud and Valérie November (2020, p.

112) explore the specific vulnerabilities of digital infrastructure control and management systems to cyber threats. Their work underlines the importance of an integrated approach to security, combining physical and digital protection of critical infrastructures.

Adapting energy infrastructures to the challenges of climate change is a major focus of this reinforcement. Sébastien Velut and Sylvain Rossiaud (2021, p. 156) analyze strategies for strengthening the resilience of power grids to extreme events. Their research highlights the importance of smart grids, diversification of energy sources and decentralized storage in increasing the flexibility and robustness of energy systems.

Finally, assessing the effectiveness of reinforcement measures is crucial for guiding future investments. Freddy Vinet and Frédéric Leone (2019, p. 289) explore methodologies for measuring the resilience of critical infrastructures to climatic hazards. Their work highlights the importance of a multi-criteria approach, integrating indicators of technical performance, service continuity and post-crisis recovery capacity.

In conclusion, strengthening critical infrastructures in the face of climatic hazards requires an integrated approach, combining technical solutions, organizational innovations and regulatory adaptations. This approach must be part of a long-term vision of regional development, combining adaptation to climate change with ecological transition. The success of these strategies hinges on close collaboration between public authorities, private operators, scientific experts and users, as well as on the ability to anticipate and continuously adapt to climate change. The major challenge is to transform climate constraints into an opportunity to rethink and modernize urban infrastructures, creating more resilient, sustainable and livable cities.

## 5.4.3 Setting up early warning systems and evacuation plans

The implementation of early warning systems and evacuation plans is a crucial element in the management of natural and technological risks in urban environments. The aim of these systems is to anticipate imminent dangers, rapidly inform the populations concerned and effectively organize their safety. The design and implementation of these systems raise complex issues, combining technical, organizational and social aspects.

Patrick Pigeon and Julien Rebotier (2019, p. 45) emphasize the importance of an integrated approach to the design of early warning systems. Their work highlights the need to consider the entire warning chain, from the detection of dangerous phenomena to the response of populations. The authors stress the importance of closely linking scientific expertise, political decision-making and public communication to ensure the effectiveness of the system.

The reliability and accuracy of detection and forecasting systems are essential for early warning. Vazken Andréassian and Charles Perrin (2020, p. 123) analyze recent advances in hydrometeorological modeling and flood forecasting. Their study highlights the importance of in situ measurement networks, satellite imagery and artificial intelligence techniques in improving the quality and speed of forecasts.

Adapting warning thresholds to local specificities is crucial to the system's effectiveness. Johnny Douvinet and Gilles Grandjean (2018, p. 89) explore approaches to defining relevant thresholds based on territorial vulnerabilities and local response capacities. Their research highlights the importance of a participatory approach, bringing together experts, decision-makers and citizens to define alert levels that are appropriate and understandable to all.

The rapid, targeted dissemination of alerts is a major challenge. Elise Beck and Samuel Rufat (2021, p. 178) analyze the different communication channels (sirens, SMS, mobile applications, social media) and their respective effectiveness according to types of risk and population profiles. Their study underlines the importance of a multi-channel strategy to reach the entire population, including vulnerable or hard-to-reach people.

Drawing up effective evacuation plans requires detailed knowledge of the territory and its dynamics. Magali Reghezza-Zitt and Alexis Sierra (2020, p. 234) explore methodologies for identifying refuge areas, sizing evacuation flows and anticipating logistical needs. Their work highlights the importance of a dynamic approach, taking into account temporal variations (day/night, week/weekend) in evacuation planning.

Taking human behavior into account in the design of warning and evacuation systems is essential. Céline Lutoff and Isabelle Ruin (2019, p. 67) analyze the factors influencing risk perception and response to warnings. Their research highlights the importance of prior awareness, trust in institutions and the adaptation of messages to different social groups in fostering an adequate response from the population.

Coordination between the various players involved in crisis management is crucial to the effectiveness of evacuations. Ludovic Fressard and Frédéric Leone (2022, p. 201) examine the ways in which government departments, local authorities, law enforcement agencies and civil protection associations work together. Their analysis highlights the importance of simulation exercises and feedback to continuously improve evacuation procedures.

The integration of new technologies into warning and evacuation systems offers promising prospects. David Provitolo and Damien Serre (2020, p. 112) explore the potential of connected objects, augmented reality and artificial intelligence to improve hazard detection, warning dissemination and evacuation guidance. Their work highlights the importance of an ethical and inclusive approach to the use of these technologies to avoid inequalities in access to information.

Adapting warning systems and evacuation plans to vulnerable people is a major challenge. Béatrice Quenault and Virginie Duvat-Magnan (2021, p. 156) analyze approaches for taking into account the specific needs of elderly, disabled or isolated people in warning and evacuation systems. Their research highlights the importance of fine-tuned vulnerability mapping and personalized support for these populations.

Finally, evaluation and continuous improvement of warning systems and evacuation plans are essential. Freddy Vinet and Anne-Catherine Chardon (2019, p. 289) explore methodologies for measuring the effectiveness of systems and identifying areas for improvement. Their work underlines the importance of post-crisis feedback and citizen involvement in evaluating and revising plans.

In conclusion, setting up effective early warning systems and evacuation plans requires a multidimensional approach, combining technical expertise, institutional organization and consideration of human and social factors. These systems must form part of an overall risk management strategy, combining prevention, preparedness and crisis management. The success of these approaches hinges on close collaboration between government departments, local authorities, scientific experts and the general public, as well as on the ability to continuously adapt systems to changing risks and technologies. The major challenge is to create warning and evacuation systems that are both technically efficient and appropriate for local populations, in order to strengthen collective resilience in the face of natural and technological hazards.

## 5.4.4 Developing local crisis and disaster management capabilities

The development of local crisis and disaster management capabilities is a crucial element in strengthening the resilience of territories in the face of natural and technological risks. This approach aims to equip local players with the skills, resources and tools they need to deal effectively with emergency situations. Implementing these strategies raises complex issues, combining training, organization and adaptation to specific local conditions.

Sandrine Revet and Julien Langumier (2019, p. 45) highlight the importance of a participatory approach to local capacity building. Their work highlights the need to involve all local stakeholders (elected representatives, technical services, associations, citizens) in the design and implementation of crisis management systems. The authors stress the importance of capitalizing on local knowledge and past experience to build a shared risk culture.

Training local players is an essential pillar of capacity building. Bruno Barroca and Damien Serre

(2020, p. 123) analyze innovative pedagogical approaches for developing crisis management skills. Their study highlights the importance of simulation exercises, serious games and feedback to promote the learning and appropriation of emergency procedures.

The development of communal safeguard plans (PCS) adapted to local realities is crucial. Johnny Douvinet and Frédéric Leone (2018, p. 89) explore methodologies for designing operational, scalable PCSs. Their research highlights the importance of an iterative approach, combining risk diagnosis, definition of roles and responsibilities, and planning of priority actions.

The establishment of volunteer networks and communal civil protection reserves offers interesting prospects for strengthening local capacities. Elise Beck and Samuel Rufat (2021, p. 178) analyze the success factors and challenges involved in mobilizing and sustaining these citizen schemes. Their study highlights the importance of institutional recognition, appropriate training and effective integration into crisis management systems.

Adapting material and logistical resources to specific local conditions is essential. Magali Reghezza-Zitt and Alexis Sierra (2020, p. 234) explore approaches for optimizing the sizing and deployment of resources according to local risks and geographical constraints. Their work highlights the importance of intercommunal pooling of resources and flexibility in their use.

The development of geographic information systems (GIS) dedicated to crisis management is a major tool for strengthening local capacities. Ludovic Fressard and Gilles Grandjean (2019, p. 67) analyze the benefits of GIS for dynamic risk mapping, locating vulnerable issues and coordinating interventions. Their research highlights the importance of training local players in the use of these tools, and of regularly updating data.

Coordination between the various territorial levels (commune, intercommunality, département, region) is crucial to effective crisis management. Céline Lutoff and Isabelle Ruin (2022, p. 201) study vertical and horizontal cooperation between institutional players. Their analysis highlights the importance of partnership agreements, inter-departmental exercises and shared feedback in improving the synergy of interventions.

The integration of new technologies into local crisis management systems offers promising prospects. David Provitolo and Damien Serre (2020, p. 112) explore the potential of drones, connected sensors and mobile applications to improve information gathering, communication and action coordination. Their work underlines the importance of a pragmatic approach, adapted to the real capacities and needs of local players.

Building community resilience is a major focus of local capacity development. Béatrice Quenault and

Virginie Duvat-Magnan (2021, p. 156) analyze approaches to mobilizing and empowering communities in the face of risk. Their research highlights the importance of self-help initiatives, solidarity networks and collective prevention actions in strengthening the social fabric and local capacity to adapt.

Finally, evaluation and continuous improvement of local crisis management systems are essential. Freddy Vinet and Anne-Catherine Chardon (2019, p. 289) explore methodologies for measuring the effectiveness of local capabilities and identifying areas for improvement. Their work underlines the importance of full-scale exercises, external audits and capitalizing on experience in order to improve practices.

In conclusion, the development of local crisis and disaster management capabilities requires a global approach, combining training, organization, equipment and citizen mobilization. This approach must be part of a local resilience strategy, combining prevention, preparedness and post-crisis management. The success of these approaches depends on the strong commitment of local elected representatives, close collaboration between public services and private players, and the active involvement of citizens. The major challenge is to create robust yet flexible systems capable of adapting to the diversity of emergency situations, while drawing on local resources and specificities. The development of local capacities thus appears to be an essential lever for strengthening the resilience of territories in the face of risks, in complementarity with national civil protection systems.

## 5.4.5 Regional and international cooperation for climate resilience

Regional and international cooperation on climate resilience has become a crucial issue, given the cross-border nature of the impacts of climate change. This collaborative approach aims to pool resources, share knowledge and coordinate actions to strengthen the adaptive capacity of territories and populations. Implementing these strategies raises complex challenges, combining environmental diplomacy, technology transfer and multi-level governance.

François Gemenne and Aleksandar Rankovic (2019, p. 45) highlight the importance of an integrated approach to international climate cooperation. Their work highlights the need to overcome North-South divides and adopt a shared vision of resilience issues. The authors emphasize the importance of exchange platforms and solidarity mechanisms for effective collective action.

Building scientific and technical capacity is an essential pillar of cooperation for climate resilience. Valérie Masson-Delmotte and Jean Jouzel (2020, p. 123) analyze international collaboration in climate research and the development of adaptation solutions. Their study highlights the importance of joint research programs, observation networks and data-sharing platforms in improving understanding of climate phenomena and their regional impacts.

Cross-border management of natural resources is a major issue for regional cooperation. Magali Reghezza-Zitt and Michel Lussault (2018, p. 89) explore approaches to coordinating the management of watersheds, shared ecosystems and water resources in a context of climate change. Their research highlights the importance of international agreements, concertation bodies and conflict resolution mechanisms for sustainable and equitable management of shared resources.

Financing adaptation to climate change is a crucial challenge for international cooperation. Nicolas Stern and Joseph Stiglitz (2021, p. 178) analyze innovative climate financing mechanisms, such as green funds, climate bonds and transnational public-private partnerships. Their study underlines the importance of an equitable approach to the distribution of financing, taking into account the specific vulnerabilities of developing countries.

The transfer of adaptation technologies and know-how is a major focus of North-South and South-South cooperation. Bruno Latour and Dipesh Chakrabarty (2020, p. 234) explore ways of disseminating and appropriating technological and organizational innovations for climate resilience. Their work highlights the importance of a contextual approach, adapting solutions to local realities and valorizing traditional knowledge.

Coordination of regional early warning and disaster management systems is crucial. Johnny Douvinet and Freddy Vinet (2019, p. 67) analyze cooperative initiatives in weather forecasting, risk monitoring and mutual assistance in the event of a crisis. Their research highlights the importance of information exchange protocols, joint exercises and cross-border solidarity mechanisms in boosting responsiveness to extreme events.

The development of regional strategies for adapting to climate change offers interesting prospects. Céline Lutoff and Isabelle Ruin (2022, p. 201) look at approaches for developing shared visions and coordinated action plans on the scale of large geographic areas (sea basins, mountain ranges, coastal zones). Their analysis highlights the importance of regional governance bodies, joint planning tools and monitoring-evaluation mechanisms in ensuring the coherence and effectiveness of actions.

Integrating climate resilience into trade agreements and development policies is an emerging issue. David Held and Angus Hervey (2020, p. 112) explore the ways in which climate, economic and social issues can be articulated in international partnerships. Their work underlines the importance of a holistic approach, integrating resilience objectives into all sectoral policies and cooperation mechanisms.

Human mobility linked to climate change raises specific challenges in terms of international cooperation. François Gemenne and Christel Cournil (2021, p. 156) analyze the legal, political and humanitarian stakes involved in population displacements induced by climate impacts. Their research highlights the importance of a concerted approach combining prevention, assistance and planned resettlement to sustainably manage these migratory flows.

Finally, strengthening institutional capacities and climate governance at national and local levels is essential for effective cooperation. Béatrice Quenault and Virginie Duvat-Magnan (2019, p. 289) explore ways of supporting developing countries in setting up policies and structures dedicated to climate resilience. Their work emphasizes the importance of experience sharing, technical assistance and capacity building in creating an environment conducive to climate action.

In conclusion, regional and international cooperation for climate resilience requires a multidimensional approach, combining political commitment, financial solidarity, scientific collaboration and citizen mobilization. This approach must be part of a shared vision of climate issues, recognizing the interdependence of nations in the face of this global challenge. The success of these approaches hinges on a balance between respect for national sovereignty and collective action, as well as on an ability to look beyond short-term interests to build sustainable solutions. The major challenge is to create a framework for cooperation that is both ambitious and operational, capable of mobilizing all the players and resources needed to strengthen climate resilience on a planetary scale.


**Conclusion:**

In conclusion, this chapter has highlighted the crucial importance of adaptation and resilience strategies in the face of the climatic and environmental challenges facing our societies. An analysis of the various approaches and measures presented reveals the complexity and multidimensionality of the issues involved in urban and territorial resilience.

We have seen that adapting to climate change requires a profound transformation in the way we plan and manage our territories. Nature-based solutions, sustainable urban planning and integrated water resource management are emerging as major ways of strengthening the resilience of our cities and ecosystems.

Reducing socio-economic vulnerabilities is also essential to building more resilient communities. This means rethinking our economic models, strengthening social protection systems, and promoting a fair and inclusive transition to more sustainable lifestyles.

Risk preparedness and crisis management are another fundamental pillar of resilience. The development of early warning systems, effective evacuation plans and the strengthening of local disaster management capacities are all crucial elements in coping with extreme events.

Last but not least, regional and international cooperation is an essential lever for meeting the global challenges of climate change. Sharing knowledge, pooling resources and coordinating actions on a supranational scale are essential to strengthening the collective resilience of our societies.

All these strategies underline the need for a holistic, integrated approach to resilience, articulating scales of action (from local to global), different sectors (environmental, economic, social), and all the players involved (public authorities, private sector, civil society).

The challenges remain considerable, both in terms of mobilizing financial and technical resources, and of changing mindsets and practices. However, the initiatives and innovations presented in this chapter show that it is possible to build more resilient territories and communities, capable of adapting and thriving in a context of global change.

Resilience thus appears not only as a necessity in the face of growing risks, but also as an opportunity to rethink our development models towards greater sustainability, equity and collective well-being. It is a major challenge for our societies, but also a promising way of building a safer, more sustainable future for present and future generations.

## *Chapter 6: Institutional capacity-building for greater resilience*

The sixth chapter of this book focuses on institutional capacity building, a crucial element in enhancing urban resilience in the face of the complex, multidimensional challenges facing contemporary cities. In a context of rapid global change and growing uncertainty, it is imperative to develop robust, adaptive governance structures capable of formulating and implementing effective urban resilience strategies.

Multilevel governance and the coordination of urban policies form the first line of analysis in this chapter. This approach recognizes that urban resilience cannot be tackled in isolation, but requires a coherent articulation of actions between different levels of government, from local to national. Emphasis is placed on clarifying roles and responsibilities, building the capacity of local authorities, and developing inter-ministerial coordination mechanisms. These elements are essential to overcome institutional silos and foster an integrated approach to urban resilience.

The second axis deals with legal and regulatory frameworks, which play a fundamental role in creating an environment conducive to urban resilience. The adaptation of urban planning and building codes, the development of specific resilience standards, and the integration of these principles into territorial planning documents are all levers for institutionalizing resilience in urban practices. The strengthening of control mechanisms and the development of legal and fiscal incentives complete this regulatory framework.

Financing urban resilience is the third area of focus. Given the scale of investment required, it is crucial to diversify funding sources and develop innovative mechanisms such as green bonds and climate funds. Strengthening cities' capacity to access international financing and setting up public-private partnerships are also being explored as promising avenues for mobilizing the resources needed to implement resilience strategies.

Finally, the chapter stresses the importance of partnerships, decentralized cooperation and exchanges of best practice. In an interconnected world, cities have every interest in pooling their experiences and drawing mutual inspiration from innovative solutions developed elsewhere. Participating in international networks of resilient cities, promoting decentralized cooperation and setting up knowledge-sharing platforms are all ways of capitalizing on collective intelligence and accelerating the spread of best practices in urban resilience.

By addressing these different aspects, this chapter offers an in-depth analysis of the institutional levers to be activated to strengthen urban resilience. It highlights the need for a systemic and multiscalar approach, involving a diversity of actors and mobilizing a variety of resources. The ultimate goal is to equip cities with the institutional capacities needed to cope with chronic shocks and stresses, while seizing opportunities to transform towards more sustainable and resilient urban development.

## 6.1 Multi-level governance and urban policy coordination

### 6.1.1 Clarification of roles and responsibilities between different levels of government

Clarifying the roles and responsibilities of the various levels of government is a fundamental challenge for effective governance of urban resilience. In the French context, characterized by a complex territorial organization and progressive decentralization, this clarification is particularly crucial to avoid overlapping competencies and optimize public action in favor of resilience.

An analysis of the distribution of powers between the State, regions, départements and municipalities reveals a complex institutional architecture that can sometimes hinder the coherent implementation of urban resilience strategies. As Patrick Le Galès (2020, p. 87) points out in his book "Le retour des villes européennes", this institutional complexity can lead to "conflicts of competence and a dilution of responsibilities, particularly damaging in the field of risk management and urban resilience".

The 2015 NOTRe law (Nouvelle Organisation Territoriale de la République) attempted to bring clarification by strengthening the role of regions and intercommunalities. However, according to Gilles Pinson (2019, p. 213) in "La ville néolibérale", "this reform has not fully resolved the problems of overlapping competences, particularly in cross-cutting areas such as urban resilience, which involve coordination between several levels of government".

There are several possible solutions to these difficulties. Firstly, the development of governance schemes clarifying the roles of each level in the implementation of resilience strategies appears to be a necessity. Romain Pasquier (2018, p. 156), in "Le pouvoir régional", recommends "setting up territorial public action conferences dedicated specifically to resilience issues, enabling vertical and horizontal coordination of policies".

In addition, strengthening the role of intercommunalities as leaders in urban resilience at the local level seems appropriate. As Alain Faure (2021, p. 78) explains in "La fabrique politique des métropoles", "intercommunalités, by virtue of their intermediate scale and their ability to federate communes, are particularly well placed to coordinate resilience actions on a catchment area scale".

Lastly, the introduction of territorial resilience contracts, inspired by State-region plan contracts, could provide a formal framework for clarifying the commitments and responsibilities of each level. According to Helluin Jean- Jacques (2022, p. 134) in "Les nouveaux défis de l'aménagement du territoire", "these contracts would make it possible to formalize resilience objectives, the resources allocated and the responsibilities of each player, while guaranteeing overall consistency".

In conclusion, the clarification of roles and responsibilities between different levels of government appears to be an essential prerequisite for strengthening the effectiveness of urban resilience policies. This clarification must be accompanied by a reflection on coordination and cooperation mechanisms between the different levels, to ensure an integrated and coherent approach to the complex challenges of urban resilience.

## 6.1.2 Setting up interministerial coordination mechanisms for urban issues

Establishing interministerial coordination mechanisms for urban issues represents a major challenge in the governance of urban resilience, particularly in the French context where administrative fragmentation is often perceived as an obstacle to coherent public action. This issue is part of a broader reflection on the transversality of public policies and the need to go beyond traditional sectoral approaches to meet the complex challenges of urban resilience.

As Pierre Muller (2018, p. 67) points out in his book "Les politiques publiques", "ministerial segmentation remains a major brake on the development of integrated policies, particularly in the urban field which, by its very nature, requires a multidimensional approach". This observation highlights the urgent need to rethink modes of inter-ministerial coordination for urban issues, and more specifically for urban resilience, which involves fields as varied as the environment, regional planning, the economy and public health.

To meet this challenge, several avenues for reflection and action can be envisaged. First of all, the creation of interministerial coordination bodies specifically dedicated to urban resilience issues appears to be a necessity. In this respect, Olivier Borraz (2020, p. 124), in "Gouverner par les risques", proposes "the creation of an interministerial committee for urban resilience, placed under the authority of the Prime Minister, to guarantee a transversal vision and the mobilization of all the administrations concerned".

Furthermore, the development of integrated strategic planning tools on a national scale could foster better coordination of departmental actions. As Gilles Pinson (2019, p. 178) explains in "La ville néolibérale", "the development of a national urban resilience plan, co-constructed by all the ministries concerned, would make it possible to define a shared vision and common objectives, thus facilitating the coordination of actions on the ground".

The introduction of cross-cutting budgeting mechanisms is also an important lever for strengthening inter-ministerial coordination. According to Michel Bouvier (2021, p. 215), in "Les finances publiques", "the adoption of interministerial budgets dedicated to urban resilience would make it possible to overcome the logic of budgetary silos and encourage collaboration between administrations".

Finally, strengthening skills and expertise within government departments in urban resilience issues appears to be an essential prerequisite for effective coordination. As Patrick Le Galès (2020, p. 156) points out in "Le retour des villes européennes", "training senior civil servants in the cross-cutting issues of urban resilience and creating interministerial expert positions dedicated to these issues would help create a common culture and facilitate exchanges between administrations".

In conclusion, setting up interministerial coordination mechanisms for urban issues, and more specifically for urban resilience, requires a profound overhaul of the way the French administration operates. This implies not only structural and organizational changes, but also a transformation of administrative cultures towards greater transversality and collaboration. This is the only way to achieve truly integrated governance of urban resilience, capable of responding to the complex challenges facing French cities.

## 6.1.3 Capacity-building for local authorities in urban management

Strengthening the urban management capabilities of local authorities is a central challenge in the quest for greater urban resilience. In the French context, marked by gradual but still incomplete decentralization, such capacity-building is a sine qua non for enabling cities to meet the complex challenges of resilience.

As Rémy Le Saout (2019, p. 87) points out in his book "Decentralization in France", "the transfer of competencies to local authorities has not always been accompanied by an equivalent transfer of resources and expertise, creating a gap between the responsibilities entrusted and the real capacities for action". This observation highlights the need for a sustained effort to equip local authorities with the human, financial and technical resources required for effective, resilient urban management.

The first step in building local authority capacity is to enhance the skills of local government employees. In this respect, Emilie Biland-Curinier (2020, p. 156), in "Les fonctionnaires territoriaux", recommends "setting up ongoing training programs specifically dedicated to the challenges of urban resilience, enabling staff to develop cutting-edge expertise in fields such as risk management, adaptation to climate change and sustainable urban planning".

Furthermore, the development of decision-support tools adapted to local realities appears to be an important lever. According to Patrick Le Galès and Gilles Pinson (2018, p. 213) in "Gouverner la ville numérique", "integrating geographic information systems (GIS) and predictive models into urban planning processes enables communities to better anticipate risks and design more effective resilience strategies".

Strengthening the financial capabilities of local authorities is also a major challenge. As Alain Guengant (2021, p. 78) explains in "Les finances locales", "increasing the fiscal autonomy of local authorities and developing more effective equalization mechanisms are necessary to enable cities, including the most modest, to finance their urban resilience projects".

Pooling resources and expertise between local authorities is a promising way of strengthening local capabilities. In "Gouverner la ville fragmentée", Hélène Reigner (2020, p. 134) points out that "the development of joint services and shared technical agencies on an inter-municipal scale enables local authorities to benefit from high-level expertise that they could not afford individually".

Finally, strengthening partnerships between local authorities and the world of research seems essential to nurture local expertise. According to Gilles Novarina (2019, p. 245) in "Plan et projet", "the creation of university chairs dedicated to urban resilience and the development of action research programs in partnership with local authorities help build bridges between academic research and urban management practices".

In conclusion, building local authority capacity in urban management requires a multi-dimensional approach, combining training, technical tools, financial resources and strategic partnerships. It is essential that local authorities develop the skills they need to play their full part in building resilient cities capable of adapting to contemporary and future challenges. This is a crucial investment for the future of our urban territories, and must be supported by an ambitious national policy of local capacity building.

## 6.1.4 Development of integrated strategic planning tools on a metropolitan scale

The development of integrated strategic planning tools on a metropolitan scale represents a crucial challenge for strengthening urban resilience in the French context. This approach aims to go beyond the limits of fragmented, sector-based planning to propose a holistic, coherent vision of urban development on the scale of large conurbations.

As Gilles Pinson (2019, p. 156) points out in his book "La ville néolibérale", "strategic metropolitan planning appears to be a necessary response to the increasing complexity of urban issues and the emergence of systemic risks that go far beyond traditional administrative boundaries". This observation highlights the urgent need to rethink our planning tools to adapt them to the reality of contemporary metropolitan dynamics.

One of the main challenges is to integrate the various dimensions of urban resilience into a coherent planning framework. In this respect, Alain Bourdin (2020, p. 89), in "L'urbanisme d'après crise", proposes "the development of master plans for metropolitan resilience, articulating environmental, social, economic and governance issues in a long-term strategic vision". These schemes would make it possible to move beyond sectoral approaches and propose an integrated roadmap for resilience on a metropolitan scale.

Taking into account the interdependencies between different urban systems is another major thrust in the development of these tools. According to Nathalie Roseau (2018, p. 213) in "Villes et infrastructures en transition", "metropolitan strategic planning must be based on a detailed analysis of the flows and networks that structure the functioning of the metropolis, in order to identify systemic vulnerabilities and design adapted resilience strategies".

The integration of different temporal scales in planning also appears to be a crucial issue. As Martin Vanier (2021, p. 134) explains in "Le pouvoir des territoires", "metropolitan strategic planning tools must make it possible to articulate short-term actions with a long-term prospective vision, while integrating mechanisms for continuous adaptation in the face of uncertainty".

In addition, the development of advanced modeling and simulation tools seems essential to feed this strategic planning. According to Patrick Le Galès and Tommaso Vitale (2020, p. 78) in "Gouverner la métropole parisienne", "the use of integrated digital models, capable of simulating the complex interactions between different urban systems, can significantly improve the quality of metropolitan planning decisions".

Citizen participation and stakeholder involvement in the planning process are also key elements. In this regard, Marie-Hélène Bacqué and Mario Gauthier (2019, p. 167), in "Participation, urbanisme et études urbaines", point out that "the development of collaborative planning tools, drawing on digital technologies and co-construction methods, makes it possible to enrich metropolitan strategic planning with the usage expertise of citizens and local stakeholders".

Finally, the articulation between metropolitan strategic planning and other scales of territorial governance appears to be a major challenge. According to Philippe Estèbe (2018, p. 245) in "L'égalité des territoires", "metropolitan planning tools must enable the coherence of local, regional and national strategies, while preserving the specificities and autonomy of each territorial level".

In conclusion, the development of integrated strategic planning tools on a metropolitan scale represents a complex but essential undertaking to strengthen urban resilience in the French context. This involves not only the development of technical methods and tools, but also a profound transformation of the urban planning culture, towards a more systemic, participative and adaptive approach. Only in this way can French metropolises equip themselves with planning frameworks that are equal to the resilience challenges they face.

## 6.1.5 Promoting participatory governance and bottom-up approaches

Promoting participatory governance and bottom-up approaches in the context of urban resilience is a major challenge for transforming traditional modes of governance and fully integrating citizens and local players in the making of the resilient city. This development is part of a broader movement to democratize public action and recognize the expertise of local residents.

As Loïc Blondiaux (2021, p. 67) points out in his book "La démocratie participative", "involving citizens in the definition and implementation of urban resilience strategies is not only a democratic imperative, but also a guarantee of the effectiveness and relevance of public policies". This observation highlights the need to rethink decision-making processes in urban planning and risk management to fully integrate the voice of citizens.

One of the main challenges is to develop participation mechanisms adapted to the complex issues of urban resilience. In this regard, Marie-Hélène Bacqué and Mario Gauthier (2019, p. 134), in "Participation, urban planning and urban studies", propose "setting up hybrid forums, bringing together experts, elected representatives and citizens, to co-build shared visions of urban resilience and collectively define action strategies". These forums for dialogue would cross-fertilize expert and lay knowledge to enrich our understanding of urban vulnerabilities and the levers of resilience.

The integration of bottom-up approaches into urban planning processes is another major thrust of this evolution. According to Gilles Pinson (2020, p. 213) in "La ville néolibérale", "collaborative planning, based on citizen initiatives and neighborhood dynamics, makes it possible to anchor resilience strategies in local realities and mobilize territories' endogenous resources". This approach implies recognizing and valorizing citizens' initiatives in the field of resilience, be they urban agriculture projects, neighborhood solidarity or social innovations.

The development of participatory digital tools also appears to be an important lever for facilitating citizen involvement. As Dominique Cardon (2018, p. 156) explains in "Digital Culture", "collaborative mapping platforms and citizen science applications offer new possibilities for involving residents in identifying risks, collecting data and proposing innovative solutions for urban resilience".

Training and capacity-building for citizens are key elements in ensuring effective participation. In this regard, Jacques Donzelot and Renaud Epstein (2019, p. 89), in "Démocratie et participation: l'exemple de la rénovation urbaine", point out that "the development of popular education programs for urban resilience and the implementation of participatory budgets dedicated to resilience projects make it possible to strengthen citizens' power to act and their ability to actively contribute to urban policies".

The articulation between participatory approaches and formal decision-making processes represents a major challenge. According to Yves Sintomer (2020, p. 178) in "Le pouvoir au peuple", "the institutionalization of citizen participation in urban planning and management procedures, notably through the creation of citizens' councils dedicated to resilience, is necessary to ensure that citizen input is genuinely taken into account in public decision-making".

Finally, the promotion of a culture of resilience at local level appears to be a crucial challenge. As Patrick Lagadec (2021, p. 245) points out in "Le temps de l'invention", "the development of participatory simulation exercises and citizen foresight approaches helps to raise awareness of resilience issues among local residents and strengthen the collective capacity to cope with crises".

In conclusion, the promotion of participatory governance and bottom-up approaches to urban resilience is an ambitious but essential task if we are to build truly resilient and inclusive cities. This evolution implies not only a transformation of institutional practices, but also a profound change in the culture of urban governance, towards a more collaborative approach open to the collective intelligence of territories. Only then will French cities be able to fully mobilize the potential of their inhabitants to meet the complex challenges of urban resilience in the 21st century.

## 6.2 Legal frameworks and regulations to promote resilience

### 6.2.1 Revision and adaptation of the town planning and building codes

The revision and adaptation of the French urban planning and building codes represent a fundamental lever for integrating the principles of urban resilience into the French regulatory framework. This approach is part of a wider need to develop urban planning standards and practices in the face of contemporary challenges, particularly those linked to climate change and emerging risks.

As Jean-Pierre Lebreton (2019, p. 45) points out in his book "Droit de l'urbanisme", "current codes, although regularly updated, still struggle to fully integrate urban resilience issues into their overall architecture. An in-depth overhaul is needed to move from a primarily regulatory approach to a more strategic and adaptive one".

First and foremost, this overhaul must take into account the evolution of scientific knowledge on urban risks. In this respect, Yves Dauge (2020, p. 112), in "La prévention des risques naturels", advocates "the systematic integration of the most recent data on climatic and environmental hazards into urban planning documents, in order to guarantee urban planning that is truly risk-informed".

Adapting building standards is another major thrust of this revision. According to Alain Maugard and Jean-Pierre Hedge (2018, p. 78) in "Regards croisés sur la construction durable", "thermal regulations must evolve towards a more global approach to the environmental performance of buildings, integrating not only energy efficiency, but also resilience to extreme climatic events and adaptability of uses".

The flexibility and adaptability of urban planning rules also appear to be crucial issues. As Elsa Vivant (2021, p. 156) explains in "L'urbanisme temporaire", "introducing transitional planning and flexible zoning into codes would enable cities to adapt more quickly to changing contexts and crisis situations, thereby strengthening their resilience".

Furthermore, the integration of circular economy principles into urban planning and building codes seems unavoidable. According to Sabine Barles (2020, p. 203) in "L'urbanisme circulaire", "the revision of codes must encourage the reuse of materials, the mutability of buildings and the design of more resilient urban infrastructures that consume fewer resources".

Last but not least, it is essential to take account of specific territorial features when adapting codes. As Gilles Novarina (2019, p. 89) points out in "Plan et projet", "the revision of codes must enable greater contextualization of standards, so as to take account of the geographical, climatic and socio-economic particularities of each territory in the definition of urban planning and construction rules".

Implementing this revision requires a collaborative approach involving a diversity of stakeholders. According to Ariella Masboungi (2018, p. 134) in "L'urbanisme négocié", "the overhaul of the codes must be based on in-depth consultation between public authorities, urban planning and construction professionals, researchers and civil society, in order to guarantee standards that are both ambitious and operational".

In conclusion, revising and adapting the French urban planning and building codes is a complex but essential task, if the principles of urban resilience are to be anchored in the French regulatory framework. This involves not only a technical updating of standards, but also a profound change in the very philosophy of these codes, towards a more integrated, flexible and adaptive approach to urban development. This is the only way to ensure that French cities have a regulatory framework that is truly conducive to building their resilience to the challenges of the 21st century.

## 6.2.2 Developing norms and standards for urban resilience

The development of norms and standards for urban resilience represents a crucial step in the formalization and operationalization of resilience principles within urban planning and management practices. The aim is to establish a common frame of reference for assessing, comparing and improving the resilience of cities to the various risks and disturbances they face.

As Serge Salat (2019, p. 78) points out in his book "Les villes résilientes", "establishing norms and standards for urban resilience is essential to move from a conceptual approach to concrete, measurable implementation of resilience principles in urban planning and management". This observation highlights the importance of standardized tools to guide the actions of decision-makers and practitioners.

One of the main challenges is to define relevant and measurable indicators of urban resilience. In this respect, Judith Rodin (2020, p. 156), in "The Resilience Dividend", proposes "the development of a composite index of urban resilience, integrating dimensions such as infrastructure robustness, economic diversity, social cohesion and institutional adaptability". This type of index would enable a holistic assessment of a city's level of resilience, and identify priority areas for improvement.

Taking account of local specificities in the development of standards is another major focus. According to Simin Davoudi (2018, p. 213) in "Resilience and the European Planning System", "urban resilience standards must be flexible enough to adapt to the varied geographical, cultural and socio-economic contexts of cities, while maintaining a common assessment framework". This approach implies developing modular standards, allowing contextualization while preserving a comparative basis.

The integration of different temporal and spatial scales in standards also appears to be a crucial issue. As David Chandler (2021, p. 134) explains in "Resilience in the Anthropocene", "urban resilience standards need to articulate short-term actions with long-term goals, and take into account the interdependencies between the scale of the neighborhood, the city and the metropolitan region".

Furthermore, the development of standardized methods for assessing urban risks and vulnerabilities seems indispensable. According to Mark Pelling (2020, p. 89) in "Urban Disaster Risk", "the establishment of standardized protocols for urban risk analysis, integrating physical, social and economic dimensions, is necessary to ensure a coherent and comprehensive approach to resilience".

Stakeholder participation in the development of standards is also a key element. In this regard, Laurent Frédéric (2019, p. 167), in "La fabrique de la ville résiliente", points out that "the involvement of local stakeholders, urban planning professionals and researchers in the standardization process helps guarantee the relevance and applicability of the standards developed".

Finally, harmonizing urban resilience standards

with other regulatory frameworks and sustainable development goals emerges as a major challenge. According to Timon McPhearson (2018, p. 245) in "Resilient Urban Futures", "resilience standards must articulate coherently with sustainable development goals, climate change mitigation policies and existing sectoral regulations".

In conclusion, the development of norms and standards for urban resilience is a complex undertaking, but one that is essential if we are to operationalize the principles of resilience in the French urban context. This involves not only technical work to define indicators and assessment methods, but also in-depth reflection on how these standards can be adapted to local specificities while maintaining a common framework. It is at this price that French cities will have standardized tools to guide and assess their efforts in the field of urban resilience, thus helping to strengthen their ability to face the challenges of the 21st century.

## 6.2.3 Integrating resilience principles into regional planning documents

The integration of resilience principles into regional planning documents represents a major challenge in terms of anchoring urban resilience in the planning and development practices of French territories. The aim is to ensure that existing planning tools, such as Schémas de Cohérence Territoriale (SCoT), Plans Locaux d'Urbanisme Intercommunaux (PLUi) and Plans Climat-Air- Énergie Territoriaux

(PCAET), systematically incorporate resilience issues across the board.

As Martin Vanier (2020, p. 112) points out in his book "Le pouvoir des territoires", "integrating resilience into planning documents must not be limited to adding an additional component, but involves a thorough overhaul of the way we conceive and organize territorial development". This observation highlights the need for a holistic and systemic approach to territorial planning.

One of the key challenges is to develop a forward-looking, adaptive vision in planning documents. In this respect, François Ascher (2019, p. 156), in "Les nouveaux principes de l'urbanisme", proposes "the introduction of long-term resilience scenarios in SCoTs, making it possible to anticipate different possible trajectories of territorial evolution in the face of risks and global change". This forward-looking approach would reinforce the ability of territories to anticipate and adapt.

Taking into account the interdependencies between different urban systems is another key aspect of this integration. According to Sabine Barles (2021, p. 213) in "L'urbanisme écologique", "planning documents must adopt a metabolic approach to the territory, analyzing material, energy and information flows to identify systemic vulnerabilities and design integrated resilience strategies".

Integrating knowledge of local risks and

vulnerabilities into planning documents also appears to be a crucial issue. As Magali Reghezza-Zitt (2018, p. 134) explains in "La ville résiliente", "PLUi must be based on in-depth diagnoses of territorial risks, integrating not only natural and technological hazards, but also social and economic vulnerabilities".

Furthermore, the development of resilience indicators adapted to planning scales seems indispensable. According to Alain Bourdin (2020, p. 89) in "L'urbanisme d'après crise", "the development of a system of territorial resilience indicators, integrated into the monitoring and evaluation tools of planning documents, would make it possible to measure progress and adjust strategies accordingly".

Citizen participation in the development of resilient planning documents is also a key element. In this regard, Marie-Christine Jaillet (2019, p. 167), in "La métropole fragile", points out that "involving residents in defining territorial resilience guidelines, particularly through participatory foresight approaches, enables planning documents to be enriched by local usage expertise and aspirations".

Finally, the articulation between different planning scales appears to be a major challenge. According to Gilles Pinson (2021, p. 245) in "La ville néolibérale", "the integration of resilience principles into planning documents must ensure the coherence of local, metropolitan and regional strategies, while preserving the specificities of each territorial level".

In conclusion, integrating the principles of resilience into territorial planning documents represents an ambitious but essential undertaking to transform the planning and development practices of French territories. This implies not only a technical evolution of documents, but also a paradigm shift in the way territorial planning is conceived, towards a more systemic, adaptive and participative approach. Only in this way will French territories be able to create planning frameworks that are truly conducive to building their resilience in the face of the challenges of the 21st century.

## 6.2.4 Strengthening control and enforcement mechanisms

Strengthening monitoring and enforcement mechanisms is a crucial element in guaranteeing the effectiveness of urban resilience norms and standards. This approach aims to ensure effective implementation of the resilience principles enshrined in urban and building codes and territorial planning documents.

As Jean-Bernard Auby (2019, p. 78) points out in his book "Droit de la ville", "the effectiveness of rules on urban resilience depends as much on the quality of the standards themselves as on the robustness of control and sanction mechanisms". This observation highlights the importance of a rigorous monitoring and enforcement system to translate regulatory ambitions into concrete realities on the ground.

One of the main challenges is to strengthen the

control capabilities of local authorities. In this respect, Patrice Duran (2020, p. 156), in "Penser l'action publique", proposes "the development of specific training for territorial agents in charge of control, in order to familiarize them with the new challenges of urban resilience and the associated evaluation methods". This skills enhancement is essential to guarantee effective and relevant control.

Another major focus is the introduction of ongoing monitoring and evaluation systems for urban projects. According to Cyria Emelianoff (2018, p. 213) in "La ville durable, une notion fossile?", "introducing regular resilience audits for major urban projects would make it possible to check compliance with standards throughout the development lifecycle and identify any necessary adjustments". This dynamic approach to control would encourage continuous adaptation to evolving risks and standards.

Involving citizens in control processes also appears to be a crucial issue. As Jacques Chevallier (2021, p. 134) explains in "L'État postmoderne", "the development of citizen vigilance mechanisms, based on participatory digital tools, can effectively complement the action of authorities in monitoring urban resilience". This approach would mobilize collective intelligence to identify non-conformities and vulnerabilities.

In addition, it seems essential to strengthen penalties for non-compliance with resilience standards. According to Marie-Anne Frison-Roche (2020, p. 89) in "Les outils de la régulation", "the introduction of dissuasive financial penalties, proportional to the potential impact on urban resilience, is necessary to encourage players to scrupulously comply with regulations". This system should be accompanied by support mechanisms to help players achieve compliance.

Coordination between the various control authorities is also a key element. In this regard, Renaud Epstein (2019, p. 167), in "La gouvernance territoriale", points out that "setting up interdepartmental platforms for sharing information and coordinating controls would optimize resources and ensure more comprehensive coverage of urban resilience issues".

Finally, adapting control mechanisms to local specificities appears to be a major challenge. According to Alain Bourdin (2018, p. 245) in "L'urbanisme d'après crise", "control mechanisms must be sufficiently flexible to adapt to varied territorial contexts, while maintaining a high level of resilience requirements".

In conclusion, strengthening control and enforcement mechanisms in the field of urban resilience is a complex but essential task if we are to guarantee the effectiveness of the norms and standards we have developed. This involves not only strengthening the resources and skills of the supervisory authorities, but also moving towards more participative and adaptive approaches to control. It is at this price that French cities will be able to ensure that the principles of resilience enshrined in their regulations are effectively translated into concrete transformations on the ground, thus helping to strengthen their ability to meet the urban challenges of the 21st century.

## 6.2.5 Developing legal and tax incentives to promote resilience

The development of legal and fiscal incentives to promote urban resilience is an important lever for encouraging public and private players to integrate resilience principles into their projects and practices. This approach aims to complement the regulatory framework with incentive mechanisms, thus creating a favorable environment for the voluntary adoption of resilience measures.

As Philippe Estèbe (2020, p. 112) points out in his book "L'égalité des territoires, une passion française", "legal and tax incentives can play a catalytic role in the spread of urban resilience practices, by aligning the economic interests of stakeholders with sustainability and adaptation objectives". This observation highlights the potential of incentives to accelerate the transition to more resilient cities.

One of the key challenges is to design effective tax incentive mechanisms. In this regard, Alain Trannoy (2019, p. 156), in "Fiscalité et développement durable", proposes "the introduction of a tax bonus-malus system for real estate projects, based on their performance in terms of resilience, integrating criteria such as adaptation to climate change, natural risk management and functional flexibility". This type of scheme could encourage developers to go beyond minimum regulatory requirements.

The introduction of innovative legal incentives is another major focus. According to Jean-Bernard Auby (2021, p. 213) in "Droit de la ville durable", "the introduction of additional building rights or accelerated authorization procedures for projects demonstrating a high level of resilience could stimulate innovation in this field". Such legal incentives would provide a tangible reward for efforts in the field of urban resilience.

Adapting incentive mechanisms to local specificities also appears to be a crucial issue. As Gilles Novarina (2018, p. 134) explains in "Plan et projet", "incentive schemes must be flexible enough to adapt to varied territorial contexts, while maintaining national consistency in the resilience objectives pursued". This approach would make it possible to take account of each territory's specific resilience challenges.

In addition, the development of incentives for the renovation of existing buildings seems essential. According to Hélène Peskine (2020, p. 89) in "La rénovation urbaine", "the introduction of tax exemption mechanisms conditional on improving the resilience of existing buildings could accelerate the upgrading of the building stock in the face of new risks". This type of incentive is particularly important in the French context, which is characterized by an ageing building stock.

Creating incentives for collective resilience initiatives is also a key element. In this regard, Cyria Emelianoff (2019, p. 167), in "La ville durable, une notion fossile?", points out that "the development of tax-advantaged participatory financing mechanisms for neighborhood-scale resilience projects could encourage collective approaches and citizen involvement".

Finally, the articulation between incentives and other public policy instruments appears to be a major challenge. According to Patrick Le Galès (2021, p. 245) in "Le retour des villes européennes", "incentive schemes must be integrated coherently into the mix of urban policy instruments, complementing regulations, public investment and awareness-raising initiatives".

In conclusion, the development of legal and fiscal incentives to promote urban resilience represents a promising area for accelerating the transition to cities better adapted to contemporary challenges. This involves not only in-depth reflection on the most effective incentive mechanisms, but also careful consideration of how they fit in with the existing regulatory framework and specific territorial features. Only in this way will French cities be able to create an environment truly conducive to the widespread adoption of urban resilience principles, mobilizing all local players in this necessary transformation.

## 6.3 Financing urban resilience and mobilizing resources

## 6.3.1 Diversifying funding sources for urban resilience projects

Diversifying sources of funding for urban resilience projects is a crucial challenge if resilience strategies are to be implemented effectively and over the long term. Given the scale of the investments required and the budgetary constraints of local authorities, it is essential to explore and mobilize a variety of innovative financial mechanisms.

As Dominique Lorrain (2018, p. 78) points out in his book "L'urbanisme 1.0", "urban resilience requires massive, long-term investments that exceed the traditional financial capacities of communities, calling for a profound overhaul of financing models". This observation highlights the urgent need to rethink approaches to financing urban resilience.

One of the main challenges is to mobilize private financing for resilience projects. In this regard, Isabelle Baraud-Serfaty (2020, p. 156), in "La nouvelle économie des territoires", proposes "the development of public-private partnerships specifically designed for resilience projects, incorporating risk and benefit sharing mechanisms adapted to the long-term nature of these investments". This approach would make it possible to raise substantial funds while aligning the interests of public and private players.

Exploiting the potential of financial markets is another major focus. According to Michel Aglietta (2019, p. 213) in "La monnaie entre dettes et souveraineté", "the issuance of green and resilient bonds by local authorities offers an opportunity to capture savings from institutional investors and individuals to finance urban resilience projects". This type of financial instrument would make it possible to diversify funding sources while raising investor awareness of resilience issues.

The development of participatory financing mechanisms also appears to be a crucial issue. As Raphaël Besson (2021, p. 134) explains in "Les Laboratoires Urbains", "crowdfunding platforms dedicated to urban resilience projects can mobilize citizen savings and strengthen local ownership of resilience initiatives". This approach would not only raise funds, but also strengthen citizen involvement in building urban resilience.

In addition, it seems essential to optimize equalization and territorial solidarity mechanisms. According to Laurent Davezies (2020, p. 89) in "L'État a toujours soutenu ses territoires", "setting up equalization funds specifically dedicated to urban resilience would make it possible to pool efforts on a national scale and support the most vulnerable territories". This type of mechanism is particularly important to ensure a fair distribution of resilience investments across the territory.

Innovative use of tax tools is also a key element. In this regard, Alain Trannoy (2019, p. 167), in "Fiscalité et développement durable", points out that "the creation of earmarked taxes or mechanisms for capturing the capital gains on property linked to resilience investments could generate dedicated, long-term resources to finance these projects".

Finally, coordination with European and international funding is a major challenge. According to Patrick Le Galès (2021, p. 245) in "Le retour des villes européennes", "French cities need to strengthen their ability to attract European funding dedicated to urban resilience, notably through better structuring of their projects and increased cooperation with other European cities".

In conclusion, diversifying sources of financing for urban resilience projects is a complex but essential task if French cities are to be effectively transformed towards greater resilience. This involves not only significant financial innovation, but also changes in governance and project management practices within local authorities. This is the only way French cities will be able to mobilize the resources they need to implement ambitious, sustainable resilience strategies that meet the challenges of the 21st century.

## 6.3.2 Development of innovative financing mechanisms (green bonds, climate funds)

The development of innovative financing mechanisms to support urban resilience projects represents a crucial challenge, given the scale of the investments required and the budgetary constraints of local authorities. Green bonds and climate funds are among the most promising tools for mobilizing capital dedicated to resilience and adaptation to climate change.

As Olivier Godard (2020, p. 112) points out in his book "La finance verte", "innovative financing mechanisms offer a unique opportunity to channel private savings into urban resilience projects, while meeting investors' growing expectations in terms of environmental and social impact". This observation highlights the potential of these tools to align financial interests with resilience objectives.

Green bonds are one of the main levers of innovative financing. In this regard, Dominique Dron (2019, p. 156), in "Finance durable et transition écologique", explains that "green bonds issued by local authorities make it possible to raise funds dedicated to specific resilience projects, while offering investors transparency on the use of funds and expected impacts". This mechanism has already been successfully adopted by several major French cities to finance climate change adaptation projects.

The development of climate funds is another major focus. According to Benoît Leguet (2021, p. 213) in "Financer la transition bas-carbone", "the creation of investment funds specialized in urban climate resilience projects makes it possible to pool risks and attract institutional investors to this emerging sector". These funds can take a variety of forms, from resilient infrastructure funds to local climate adaptation funds.

Innovation in financing structures also appears to be a crucial issue. As Michel Aglietta (2018, p. 134) explains in "Money between debt and sovereignty", "the development of green securitization structures, making it possible to pool and refinance portfolios of urban resilience projects, could significantly increase the financing capacity of communities". This approach would make it possible to mobilize capital on a large scale, while reducing the risks perceived by investors.

The use of results-based financing mechanisms also seems promising. According to Emilie Alberola (2020, p. 89) in "Les instruments économiques au service du climat", "Environmental Impact Bonds offer an innovative way of financing resilience projects by linking financial returns to the achievement of measurable urban resilience objectives". This type of instrument enables risks to be shared between the public and private sectors, while guaranteeing investment efficiency.

The tokenization of resilience assets also represents an innovative avenue. In this regard, Pierre-Jean Benghozi (2019, p. 167), in "Blockchain and sustainable finance", points out that "using blockchain technology to create tokens representing shares in urban resilience projects could democratize access to these investments and facilitate the mobilization of citizen savings".

Finally, the articulation between these innovative mechanisms and traditional financing tools appears to be a major challenge. According to Patrick Artus (2021, p. 245) in "Et si les salariés se révoltaient?", "integrating resilience criteria into financial rating mechanisms and stock market indices could enhance the attractiveness of urban resilience projects to mainstream investors".

In conclusion, the development of innovative financing mechanisms such as green bonds and climate funds represents a crucial undertaking to increase the capacity of French cities to finance their transformation towards greater resilience. This involves not only significant financial innovation, but also changes in project management and reporting practices within local authorities. It is at this price that cities will be able to mobilize the capital needed to implement ambitious resilience strategies, while contributing to the emergence of sustainable finance aligned with the challenges of the 21st century.

### 6.3.3 Strengthening cities' ability to access international financing

Strengthening the capacity of French cities to access international funding dedicated to urban resilience is a major challenge if we are to amplify the resources available for action. Faced with the growing number of international funds dedicated to climate change adaptation and resilience, it is crucial that local authorities develop their expertise and strategies to capture these resources.

As Harriet Bulkeley (2019, p. 78) points out in

her book "Cities and Climate Change", "access to international funding represents not only an opportunity for additional resources for cities, but also a means of becoming part of global networks of expertise and innovation in urban resilience". This observation highlights the multiple benefits that French cities can derive from a better connection to international financing circuits.

One of the main challenges is to strengthen local authorities' in-house expertise in international financial engineering. In this respect, Emmanuelle Baudoin (2020, p. 156), in "L'action internationale des collectivités territoriales", proposes "the creation of dedicated units within municipal departments, staffed by experts in international financing and in setting up resilience projects, capable of navigating the complexity of calls for projects and application procedures". This professionalization is essential to optimize the chances of success in obtaining international funding.

The development of strategic partnerships is another major focus. According to Eric Huybrechts (2021, p. 213) in "Les réseaux de villes face aux défis globaux", "the establishment of consortia between French and European cities makes it possible to pool resources and expertise to respond to large-scale international calls for projects". This collaborative approach strengthens the credibility of bids and facilitates the sharing of experience.

Improving the international visibility of local resilience initiatives also appears to be a crucial issue. As Michèle Pappalardo (2018, p. 134) explains in "Diplomatie des villes", "developing an international communication strategy on urban resilience projects, relying in particular on social networks and specialized platforms, can draw the attention of international donors to innovative local initiatives". This approach is essential if projects are to stand out from the crowd in a context of heightened competition for funding.

In addition, strengthening ties with international financial institutions seems indispensable. According to Michel Aglietta (2020, p. 89) in "La monnaie entre dettes et souveraineté", "establishing direct, ongoing relations with institutions such as the European Investment Bank or the World Bank can facilitate French cities' access to specific lines of financing for urban resilience". This type of institutional partnership can open up long-term financing opportunities.

Adapting projects to the criteria of international funding is also a key element. In this regard, Jean-Pierre Elong Mbassi (2019, p. 167), Secretary General of UCLG Africa, points out in a report that "cities must learn to formulate their resilience projects in such a way as to respond explicitly to the objectives and indicators of international funds, while preserving their local relevance".

Lastly, developing an integrated approach to resilience appears to be a major challenge for accessing international funding. According to Laurence Tubiana (2021, p. 245), CEO of the European Climate Foundation, in an interview for "Le Monde", "the cities that succeed in attracting international funding are those that present a holistic vision of resilience, integrating environmental, social and economic dimensions into their projects".

In conclusion, building the capacity of French cities to access international funding for urban resilience is a crucial task if the available means of action are to be amplified. This involves not only developing skills and strategies at local level, but also changing the way resilience projects are designed and presented. Only in this way will French cities be able to take full advantage of the opportunities offered by international funding, strengthening their ability to implement ambitious and innovative resilience strategies.

## 6.3.4 Setting up public-private partnerships to finance resilience

Setting up public-private partnerships (PPPs) to finance urban resilience represents an innovative and potentially fruitful approach to mobilizing additional resources and sharing the risks associated with investments in this field. Given the scale of the challenges and the budgetary constraints faced by communities, PPPs offer a promising way of accelerating the implementation of ambitious resilience projects.

As Frédéric Marty (2020, p. 112) points out in his book "Les partenariats public-privé", "PPPs adapted to the challenges of urban resilience make it possible not only to mobilize private capital, but also to integrate the expertise and innovation of the private sector into the design and management of resilient infrastructures". This observation highlights the dual financial and technical advantages of PPPs in the context of urban resilience.

One of the main challenges is to design PPP models specifically adapted to resilience projects. In this regard, Isabelle Baraud-Serfaty (2019, p. 156), in "La nouvelle économie des territoires", proposes "the development of resilience performance contracts, where the private partner's remuneration is partly linked to the achievement of measurable objectives in terms of urban resilience". This type of mechanism would align the interests of public and private players with long-term resilience objectives.

The integration of positive externalities into PPP economic models is another major focus. According to Dominique Bureau (2021, p. 213) in "Économie des partenariats public-privé", "valuing the co-benefits of resilience projects, such as improved public health or local job creation, can enhance the economic attractiveness of PPPs in this field". This approach would make it possible to justify larger investments by taking into account all the positive spin-offs for the territory.

The development of innovative risk-sharing mechanisms also appears to be a crucial issue. As Stéphane Saussier (2018, p. 134) explains in "The Economics of Public Contracts", "the use of parametric insurance mechanisms or targeted public guarantees can facilitate private sector engagement in resilience projects involving significant climate risks". These tools would reduce the financial exposure of private partners, while maintaining their incentive to perform.

The involvement of local players in PPPs also seems essential. According to Raphaël Besson (2020, p. 89) in "Les Laboratoires Urbains", "integrating local businesses and civil society organizations into PPP consortiums can strengthen the territorial anchoring of resilience projects and facilitate their social acceptability". This approach would make it possible to combine global expertise with detailed knowledge of the local context.

Setting up appropriate governance structures is also a key element. In this regard, Jean-Pierre Gaudin (2019, p. 167), in "Critique de la gouvernance", points out that "the creation of multi-stakeholder steering committees, integrating public, private and citizen representatives, can improve the transparency and effectiveness of PPPs in the field of urban resilience".

Finally, the articulation between PPPs and other resilience financing tools appears to be a major challenge. According to Michel Aglietta (2021, p. 245) in "La monnaie entre dettes et souveraineté", "PPPs must be coherently integrated into a financing mix that also includes green bonds, dedicated funds and traditional public financing, in order to optimize the overall financing structure of urban resilience strategies".

In conclusion, setting up public-private partnerships to finance urban resilience represents a promising way of accelerating the transformation of French cities in the face of contemporary challenges. This approach implies not only innovation in the structuring of contracts and financial mechanisms, but also an evolution in governance and the very design of resilience projects. Only then will PPPs be able to make their full contribution to mobilizing the resources and expertise needed to implement ambitious, long-term resilience strategies, while ensuring a balanced sharing of risks and rewards between public and private players.

## 6.3.5 Integrating resilience costs into long-term budget planning

Integrating the costs of resilience into the long-term budgetary planning of French cities is a fundamental challenge if urban resilience strategies are to be sustainable and effective. This implies a paradigm shift in the financial management of local authorities, from a reactive approach to a proactive, anticipatory vision of the investments needed to meet future challenges.

As Alain Trannoy (2020, p. 78) points out in his book "Économie des finances publiques", "the systematic integration of resilience costs into municipal budgets requires a profound overhaul of financial planning methods, incorporating longer time horizons and multiple scenarios linked to climate and social risks". This observation highlights the complexity and scale of the change required in current budgeting practices.

One of the main challenges is to develop economic evaluation tools adapted to resilience projects. In this regard, Émilie Coudel (2019, p. 156), in "L'évaluation des politiques de développement durable", proposes "the adoption of cost-benefit analysis methods incorporating the long-term positive externalities of resilience investments, such as reduced future disaster costs or improved quality of life". This approach would enable resilience investments to be better justified and prioritized in budgetary trade-offs.

Setting up financial provisions dedicated to resilience is another major thrust. According to Michel Klopfer (2021, p. 213) in "La gestion financière des collectivités locales", "the creation of reserve funds specifically allocated to resilience investments, fed by a fixed share of the annual budget, can guarantee the availability of resources for long-term projects, regardless of short-term budget fluctuations". This mechanism would make it possible to safeguard funding for resilience, ensuring continuity of action.

The development of green and resilient accounting also appears to be a crucial issue. As

Jacques Richard (2018, p. 134) explains in "Accounting for the 21st Century", "integrating resilience indicators into the accounting and financial documents of local authorities would make it possible to better reflect the real value of urban assets in the face of future risks and guide investment decisions". This accounting evolution would contribute to a better consideration of resilience issues in the asset management of cities.

The link between urban planning and financial planning also seems essential. According to Patrizia Ingallina (2020, p. 89) in "Le projet urbain", "the systematic integration of a long-term financial dimension into urban planning documents, in particular PLUs and SCOTs, would ensure consistency between resilience ambitions and the financing capacities of local authorities". This approach would facilitate the anticipation and programming of necessary long-term investments.

The introduction of financial smoothing mechanisms is also a key element. In this regard, Gilbert Cette (2019, p. 167), in "Le bilan des réformes Macron", points out that "the use of financial instruments such as very long-term bond issues or green securitization mechanisms can spread over time the cost of the massive investments required for urban resilience".

Finally, the development of a culture of evaluation and continuous adjustment appears to be a major challenge. According to Patrice Duran (2021, p.

245) in "L'évaluation des politiques publiques", "the establishment of regular processes for reviewing and adjusting resilience-related financial plans, based on rigorous impact assessments, is essential to maintain the effectiveness and relevance of investments over the long term".

In conclusion, integrating the costs of resilience into the long-term budgetary planning of French cities is a complex but essential task if urban resilience strategies are to be implemented effectively and sustainably. This involves not only a change in financial management tools and practices, but also a profound change in the way public investments are conceived and valued. Only then will cities be able to align their financial resources with their resilience ambitions, guaranteeing their ability to meet future challenges while optimizing the use of public funds over the long term.

## 6.4 Partnerships, decentralized cooperation and exchanges of best practices

## 6.4.1 Development of national and international networks of resilient cities

The development of networks of resilient cities, both nationally and internationally, represents a major strategic lever for accelerating the transition to more robust and adaptable urban models. These networks offer invaluable platforms for exchanging experience, pooling resources and co-constructing innovative solutions to the common challenges of urban resilience.

As Michele Acuto (2019, p. 45) points out in her book "Global City Challenges", "networks of resilient cities act as innovation gas pedals, enabling the rapid dissemination of best practices and a rise in collective competence on resilience issues". This observation highlights the catalytic role of these networks in transforming urban practices.

At national level, the creation of specialized thematic networks appears to be an appropriate approach. In this respect, Cyria Emelianoff (2020, p. 112), in "La ville durable, une notion à géométrie variable", suggests "setting up communities of practice between French cities, structured around specific issues such as flood risk management or adaptation to urban heat". This approach would enable targeted sharing of expertise and feedback between communities facing similar challenges.

Linking national and international networks is another major focus. According to Harriet Bulkeley (2021, p. 78) in "Cities and Climate Change", "the interconnection between local networks and global initiatives such as the 100 Resilient Cities program enables French cities to benefit from international expertise while promoting their own innovations on a global scale". This multi-scale approach strengthens the capacity for action and visibility of cities committed to resilience.

The development of digital collaborative tools also appears to be a crucial issue. As Antoine Picon

(2018, p. 156) explains in "Smart Cities", "the creation of digital platforms dedicated to urban resilience, integrating functionalities for data sharing, collaborative modeling and connecting actors, can significantly amplify the impact of city networks". These tools would encourage smoother collaboration and effective capitalization of knowledge within networks.

In addition, the involvement of non-state actors in these networks seems indispensable. According to Saskia Sassen (2020, p. 89) in "The Global City", "integrating businesses, universities and civil society organizations into networks of resilient cities enriches exchanges and fosters the emergence of innovative partnerships". This multi-stakeholder approach would strengthen the capacity to innovate and implement resilience solutions.

The establishment of collaborative financing mechanisms is also a key element. In this regard, Olivier Coutard (2019, p. 167), in "Urban Energy Transitions", points out that "the creation of pooled investment funds between member cities of the same network can facilitate the financing of pilot resilience projects and accelerate the scaling-up of promising solutions".

Finally, the development of city diplomacy focused on resilience appears to be a major challenge. According to Benjamin Barber (2021, p. 245) in "If Mayors Ruled the World", "strengthening the diplomatic capacities of cities within international networks enables them to carry more weight in global negotiations on climate and resilience, thus complementing the action of states".

In conclusion, the development of networks of resilient cities at national and international level represents an essential strategic lever for accelerating the transition to more robust and adaptable urban models. This involves not only the creation of structures for exchange and collaboration, but also the development of tools, financing mechanisms and specific skills. It is at this price that French cities will be able to benefit fully from the collective dynamics of resilience networks, strengthening their capacity to innovate, learn and implement ambitious and effective resilience strategies in the face of the challenges of the 21st century.

## 6.4.2 Promoting decentralized cooperation and twinning focused on resilience

The promotion of decentralized cooperation and twinning focused on resilience represents a significant opportunity for French cities to strengthen their capacities and enrich their approaches to urban resilience. This form of international cooperation, rooted in direct relations between local authorities, offers a framework conducive to the exchange of experience, the transfer of skills and the implementation of innovative joint projects.

As Bertrand Gallet (2019, p. 67) points out in his

book "La coopération décentralisée des collectivités territoriales", "city-to-city partnerships focused on resilience enable not only mutual learning, but also the joint mobilization of resources and expertise to address common challenges". This observation highlights the transformative potential of these international collaborations.

One of the main challenges is to identify and structure relevant resilience partnerships. In this respect, Marie-Christine Jaillet (2020, p. 123), in "Les villes en partage", suggests "establishing specific criteria for choosing partner cities, based on the complementarity of resilience expertise and the similarity of climate and social challenges". This approach would maximize the added value of exchanges and collaborations.

Another key area is the development of staff exchange programs. According to Eric Huybrechts (2021, p. 189) in "Les réseaux de villes face aux défis globaux", "setting up professional mobility programs between technical departments of partner cities can accelerate the transfer of skills and the adoption of new urban resilience practices". These exchanges would encourage cross-fertilization of approaches and know-how.

The development of joint resilience projects also appears to be a crucial issue. As Michèle Pappalardo (2018, p. 145) explains in "Diplomatie des villes", "the co-construction of pilot resilience projects between twinned cities, with shared risks and benefits, can stimulate innovation and facilitate access to international funding". This collaborative approach would help pool resources and amplify the impact of resilience initiatives.

In addition, the involvement of local players in partnerships seems essential. According to Yves Viltard (2020, p. 78) in "L'action internationale des collectivités territoriales", "integrating local businesses, universities and associations into decentralized cooperation projects on resilience strengthens the territorial anchoring and sustainability of initiatives". This multi-stakeholder approach would help to ensure that resilience issues are widely taken on board by communities.

Setting up appropriate monitoring and evaluation mechanisms is also a key element. In this regard, Franck Petiteville (2019, p. 201), in "La coopération décentralisée", points out that "establishing common resilience indicators and carrying out regular joint evaluations make it possible to measure the real impact of partnerships and adjust cooperation strategies".

Finally, the development of effective communication around cooperation initiatives appears to be a major challenge. According to Anne-Marie Autissier (2021, p. 167) in "L'Europe des festivals", "the media promotion of resilience projects resulting from twinning can strengthen citizen and political commitment to these partnerships, while inspiring other communities".

In conclusion, the promotion of decentralized cooperation and twinning focused on resilience represents a promising way of strengthening the capacities of French towns and cities in the face of contemporary challenges. This involves not only structuring relevant partnerships and implementing joint projects, but also developing specific skills in international cooperation. Only in this way will cities be able to fully benefit from the transformative potential of these collaborations, enriching their resilience approaches through fruitful international dialogue and contributing to the emergence of a global community of practice in urban resilience.

### 6.4.3 Participation in global initiatives on urban resilience (100 Resilient Cities, etc.)

The participation of French cities in global initiatives on urban resilience, such as the 100 Resilient Cities (100RC) program initiated by the Rockefeller Foundation, represents a major opportunity to accelerate their transition to more resilient urban models. These international platforms offer a structuring framework, significant resources and a global network of experts to support cities in developing and implementing their resilience strategies.

As Michael Berkowitz (2019, p. 34), former president of 100RC, points out in "Building Urban Resilience", "global initiatives like 100RC act as catalysts, enabling cities to benefit from cutting-edge expertise, financial support and an international network to propel their resilience efforts." This observation highlights the transformative potential of these programs for participating cities.

One of the main advantages of these initiatives is access to a proven methodology. In this respect, Judith Rodin (2020, p. 156), in "The Resilience Dividend", explains that "the methodological framework developed by 100RC, structured around systemic risk assessment and the development of holistic strategies, offers cities a robust approach to tackling the complexity of resilience issues". This methodology provides cities with a structured and proven process for developing their strategies.

The funding of a Chief Resilience Officer (CRO) position is another major asset of these programs. According to Arnoldo Matus Kramer (2021, p. 78), former CRO of Mexico City, in "Urban Resilience in Action", "the creation of a dedicated resilience position within the municipal administration, supported by international expertise, helps catalyze cross-cutting efforts and maintain resilience as a strategic priority". This dedicated leadership plays a crucial role in coordinating and driving forward resilience initiatives.

Access to a global network of experts and practitioners also appears to be a crucial benefit. As Lauren Sorkin (2018, p. 123), Director of the Global Resilient Cities Network, explains in "Networked

Urban Resilience", "participation in these initiatives offers cities privileged access to an international community of practice, facilitating the sharing of experience and the identification of innovative solutions". This network enables cities to inspire each other and accelerate their learning.

In addition, the international visibility offered by these programs seems to be a significant asset. According to Saskia Sassen (2020, p. 89) in "The Global City", "integration into global initiatives such as 100RC strengthens the positioning of cities on the international stage, potentially attracting resilience-related investments and partnerships". This visibility can translate into increased opportunities for funding and collaboration.

Connecting with private sector partners is also a key element of these initiatives. In this regard, Peter Williams (2019, p. 201), in "Resilient Cities", points out that "programs like 100RC facilitate dialogue between cities and innovative resilience companies, paving the way for innovative public-private partnerships".

Finally, access to technological tools and platforms appears to be a significant advantage. According to Anthony Townsend (2021, p. 167) in "Smart Cities", "global resilience initiatives often offer participating cities access to advanced risk modeling and analysis tools, enhancing their ability to make informed decisions".

In conclusion, the participation of French cities in global initiatives on urban resilience represents a strategic opportunity to accelerate and strengthen their resilience initiatives. This participation offers not only a robust methodological framework and dedicated resources, but also access to a global network of expertise and innovation. However, to take full advantage of these opportunities, cities must ensure that the contributions of these initiatives are effectively integrated into their existing structures and processes, while adapting global approaches to their specific local contexts. Only in this way can participation in these international programs truly catalyze the transformation of French cities towards more resilient models, capable of meeting the complex challenges of the 21st century.

### 6.4.4 Setting up platforms for exchanging knowledge and best practices

Setting up platforms for exchanging knowledge and best practices is an essential lever for accelerating and amplifying urban resilience efforts in French cities. These platforms offer a structured space for capitalizing on, sharing and disseminating experience and innovations in resilience, thus promoting collective learning and a generalized increase in the skills of urban players.

As Etienne Wenger (2018, p. 45) points out in his book "Communities of Practice", "knowledge exchange platforms act as catalysts for innovation, enabling the cross-fertilization of ideas and accelerating the adoption of best practices". This observation highlights the transformative potential of these collaborative tools for urban resilience.

One of the main challenges is to design platforms tailored to the specific needs of urban resilience stakeholders. In this respect, Saskia Sassen (2020, p. 112), in "The Global City", proposes "the creation of thematic platforms structured around the main challenges of urban resilience, such as climate adaptation, risk management or social cohesion". This approach would enable targeted and relevant sharing of knowledge between actors facing similar challenges.

The development of tools for capitalizing on and formalizing knowledge is another major focus. According to Ikujiro Nonaka (2021, p. 78) in "Knowledge Creation in Urban Planning", "the integration of systematic experience capitalization methodologies, such as structured feedback or in-depth case studies, is essential to transform individual practices into shareable knowledge". These methods would make it possible to effectively capitalize on the learnings from resilience initiatives.

The active animation of user communities also appears to be a crucial issue. As Jane Jacobs (2019, p. 156) explains in "The Death and Life of Great American Cities", "the vitality of exchange platforms depends heavily on the ongoing animation of user communities, through the organization of virtual events, webinars or innovation challenges". Such animation would foster a dynamic of exchange and mutual learning between participants.

In addition, the integration of co-creation functionalities seems indispensable. According to Henry Chesbrough (2020, p. 89) in "Open Innovation", "the inclusion of online collaborative workspaces, enabling users to co-construct solutions or collectively solve problems, can significantly enrich the value of exchange platforms". These features would encourage the emergence of innovative solutions based on collective intelligence.

Establishing mechanisms for validating and certifying best practices is also a key element. In this regard, Elinor Ostrom (2019, p. 167), in "Governing the Commons", points out that "establishing peer-review and labeling processes for best practices can strengthen the credibility and dissemination of knowledge shared on platforms".

Finally, the development of interfaces with city information systems appears to be a major challenge. According to Carlo Ratti (2021, p. 245) in "The City of Tomorrow", "the integration of exchange platforms with urban management systems and city modeling tools can facilitate the concrete application of shared knowledge in decision-making and planning processes".

In conclusion, setting up platforms for the exchange of knowledge and best practices represents an essential strategic lever for accelerating and amplifying urban resilience efforts in French cities. This requires not only the development of appropriate technological tools, but also the implementation of processes for capitalizing on, coordinating and validating knowledge. Only in this way can these platforms truly play their role as catalysts for innovation and gas pedals of the transition to more resilient urban models.

To maximize the impact of these platforms, it is crucial to adopt an inclusive approach, involving not only institutional players but also citizens, businesses and civil society organizations. In addition, particular attention needs to be paid to the accessibility and ergonomics of interfaces, to ensure widespread and effective use by all urban resilience stakeholders.

Finally, it is important to consider these platforms not as ends in themselves, but as tools at the service of a broader dynamic of collective learning and innovation. Their success will depend on their ability to integrate harmoniously into existing urban management and planning ecosystems, while catalyzing new forms of collaboration and collective intelligence in the service of the resilience of French cities.

## 6.4.5 Strengthening partnerships with the academic world and research centers

Strengthening partnerships between cities and the academic world, including universities and research centers, is a major strategic lever for improving urban resilience. These collaborations make it possible to mobilize cutting-edge knowledge, develop innovative approaches and anchor resilience practices in a rigorous scientific approach.

As Michael Batty (2018, p. 23) points out in his book Inventing Future Cities, "the interface between academic research and urban practice is fertile ground for innovation in resilience, enabling theoretical models to be confronted with realities on the ground". This observation highlights the transformative potential of these partnerships to rethink and strengthen city resilience.

One of the main challenges is to structure effective and sustainable collaborations. In this respect, Sheila Jasanoff (2020, p. 112), in "Science and Public Reason", proposes "the establishment of shared governance structures between cities and academic institutions, such as joint urban laboratories or research chairs dedicated to resilience". Such arrangements would institutionalize collaboration and guarantee long-term commitment from stakeholders.

The development of action-research programs is another major focus. According to Yvonne Rydin (2021, p. 78) in "The Future of Planning", "the development of co-constructed research projects between researchers and urban practitioners, anchored in concrete resilience issues, promotes the production of directly applicable knowledge". This collaborative approach bridges the gap between theory and practice.

The integration of researchers into municipal services also appears to be a crucial issue. As Patrick Le Galès (2019, p. 156) explains in "European Cities", "the presence of researchers in residence in urban administrations can facilitate the transfer of knowledge and the adoption of scientific methodologies in decision-making processes linked to resilience". This immersion would encourage cross-fertilization between research and public action.

Ongoing training for city professionals also seems essential. According to Susan Fainstein (2020, p. 89) in "The Just City", "the introduction of continuing education programs co-developed by universities and cities helps keep practitioners' skills up to date on the latest advances in urban resilience". Such training would strengthen cities' ability to integrate innovations into their practices.

The creation of spaces for dialogue and exchange is also a key element. In this regard, Richard Sennett (2019, p. 167), in "Building and Dwelling", points out that "the regular organization of hybrid forums, bringing together researchers, practitioners and citizens around resilience issues, can stimulate the emergence of innovative ideas and foster the collective appropriation of scientific knowledge".

Finally, the development of open data platforms appears to be a major challenge. According to Rob Kitchin (2021, p. 245) in "The Data Revolution", "the establishment of systems for sharing urban data between cities and researchers, in compliance with ethical and confidentiality rules, can considerably enrich resilience research and improve evidence-based decision-making".

In conclusion, strengthening partnerships with the academic world and research centers represents a major opportunity for French cities to improve their resilience. This involves not only the structuring of institutional collaborations, but also the development of new practices for the co-production of knowledge and the transfer of expertise.

To maximize the impact of these partnerships, it is crucial to adopt a transdisciplinary approach, mobilizing a wide range of expertise from the natural sciences to the social sciences. In addition, particular attention must be paid to translating research findings into tools and recommendations that can be directly applied by urban decision-makers.

It is also important to see these partnerships as catalysts for social innovation, encouraging experimentation and the adoption of new approaches to governance and urban planning. Their success will depend on their ability to build lasting bridges between the worlds of research and public action, while remaining rooted in the realities and specific needs of the territories concerned.

Last but not least, these collaborations must be part of an international perspective, encouraging exchanges and comparisons between cities in different countries. This global dimension would not only enrich local approaches, but also position French cities as leading players in urban resilience research and innovation on a global scale.

## Conclusion:

Strengthening cooperation and partnerships appears to be an essential lever for accelerating and amplifying the resilience efforts of French cities. As Saskia Sassen (2021, p. 312) points out in "The Global City", "the complexity of contemporary urban challenges demands a collaborative, multiscalar approach, transcending traditional institutional and geographical boundaries".

The various sections of this chapter have highlighted several strategic axes:

1. Strengthening interdepartmental and inter-municipal coordination is a sine qua non for developing a holistic approach to resilience. As Patrick Le Galès (2020, p. 178) argues in "European Cities", "urban resilience can only be conceived through integrated governance, overcoming administrative silos".

2. The promotion of decentralized cooperation and twinning focused on resilience offers French towns and cities

   opportunities for mutual learning and shared innovation. According to Bertrand Badie (2019, p. 45) in "La diplomatie des villes", "these international partnerships are veritable experimental laboratories for urban resilience".

3. Participation in global initiatives on urban resilience, such as 100 Resilient Cities, gives cities access to global resources, expertise and networks. Eric Corijn (2021, p. 89) in "The Urbanity of Science" points out that "these international platforms act as catalysts for innovation and the acceleration of urban transitions".

4. Setting up platforms for exchanging knowledge and best practices encourages the capitalization and dissemination of successful experiences. According to Etienne Wenger (2018, p. 156) in "Communities of Practice", "these collaborative spaces are essential for building a collective intelligence of urban resilience".

5. Strengthening partnerships with the academic world and research centers helps to anchor resilience strategies in a rigorous scientific approach. Michael Batty (2020, p. 234) in "Inventing Future Cities" asserts that "the interface between research and practice is the crucible of urban innovation".

These different forms of cooperation and partnership are not mutually exclusive, but rather complementary. Their coherent articulation can create an ecosystem conducive to the emergence of innovative solutions and the acceleration of the urban transformations needed to meet the challenges of the 21st century.

However, the effective implementation of these collaborations raises a number of challenges. As Ash Amin (2021, p. 278) notes in "Seeing Like a City", "moving from a culture of competition to one of cooperation between cities and institutions requires a profound shift in paradigm and practice".

For the future, it is crucial to:

1. Develop institutional and legal frameworks that facilitate multi-stakeholder, multi-level cooperation.

2. Strengthen the skills of urban players in managing complex partnerships.

3. Set up innovative financing mechanisms to support these collaborations over the long term.

4. Systematically integrate a collaborative dimension into the development and implementation of urban resilience strategies.

In conclusion, strengthening cooperation and partnerships appears to be a sine qua non for building resilient cities, capable of adapting and transforming in the face of the complex, interconnected challenges of the contemporary world. As Richard Sennett (2018, p. 301) states in "Building and Dwelling", "the future of our cities will depend on our ability to build creative and sustainable alliances, transcending traditional boundaries between disciplines, sectors and territories".

## *General conclusion: Towards more resilient Moroccan cities; Prospects and recommendations*

The in-depth analysis of urban resilience in the Moroccan context highlighted the multiple challenges faced by the country's cities in the face of globalization and global change. The theoretical conceptualization of resilience as a multidimensional paradigm proved particularly relevant to understanding the complexity of contemporary urban dynamics. Indeed, the holistic approach adopted in this study, integrating the economic, social, environmental and institutional dimensions of resilience, offers a robust analytical framework for assessing the capacity of Moroccan cities to adapt and transform.

An examination of the impact of economic globalization on Moroccan urban areas has highlighted the structural vulnerabilities inherited from a development model long based on dependent industrialization and uneven tertiarization of the economic fabric. The economic restructuring induced by international openness has led to deindustrialization and the casualization of urban employment, exacerbating socio-spatial inequalities and demographic pressures linked to rural exodus. Against this backdrop, strengthening the economic resilience of Moroccan cities is a strategic imperative, requiring proactive policies to diversify production, attract investment and support local entrepreneurship and the social economy.

The social dimension of urban resilience is crucial in tackling the challenges of poverty, exclusion and community tensions that threaten the cohesion of Moroccan cities. Improving access to essential services, strengthening citizen participation and promoting cultural heritage as a vector of local identity are priority areas for building more cohesive and resilient urban communities. Urban security and conflict prevention are also emerging as major challenges, calling for integrated approaches combining urban planning, social cohesion and inclusive governance.

In the face of urgent climate change and growing environmental risks, the ecological resilience of Moroccan cities is a fundamental pillar of their sustainable development. The adoption of ambitious climate change adaptation strategies, coupled with proactive energy transition and urban ecosystem preservation policies, is essential. Resilient urban planning, integrating the principles of sustainable urbanism and disaster risk reduction, is an essential lever for shaping more environmentally resilient Moroccan cities.

Institutional capacity-building is emerging as a sine qua non condition for the effective operationalization of urban resilience strategies in Morocco. Improving multi-level governance, establishing appropriate legal and regulatory frameworks, and developing innovative resilience financing mechanisms are all priorities for public decision-makers. Promoting multi-stakeholder partnerships and strengthening decentralized cooperation also offer promising opportunities for pooling resources and capitalizing on best practices in urban resilience.

Ultimately, this study highlights the need for a systemic and integrated approach to urban resilience in Morocco, transcending traditional sectoral silos. Building more resilient Moroccan cities requires a long-term strategic vision, articulating interventions at different spatial and temporal scales. It also implies a paradigm shift in urban governance, encouraging experimentation, collective learning and continuous adaptation in the face of a constantly changing environment. While significant progress has been made in recent years, sustained efforts will be needed to firmly anchor resilience at the heart of Moroccan urban policies and meet the complex challenges of the 21st century.

## *List of bibliographies :*

- Agénor, P. R. and El Aynaoui, K. (2015). Morocco: Growth strategy to 2025 in a changing international environment. Rabat: OCP Policy Center.
- Ameur, M. (2016). New urban territories in Morocco: Between globalization and local development. Rabat: Éditions universitaires du Maroc.
- Barthel, P. A. and Planel, S. (2010). Tanger-Med and Casa- Marina, prestige projects in Morocco: New capitalist frameworks and local context. Environnement Bâti, 36(2), 176-191.
- Belkadi, A. (2018). City policy and social cohesion in Morocco. Casablanca: Éditions Afrique Orient.
- Benlahcen Tlemçani, M. (2018). Urban governance and sustainable development in Morocco. Casablanca: Éditions La Croisée des Chemins.
- Berriane, M. and Signoles, P. (eds.) (2015). Les terroirs au Sud, vers un nouveau modèle? : Une expérience marocaine. Paris: Karthala Editions.
- Bogaert, K. (2018). Globalized authoritarianism: Megaprojects, slums, and class relations in urban Morocco. Minneapolis: University of Minnesota Press.
- Catusse, M. and Vairel, F. (eds.) (2016). Morocco in the present: D'une époque à l'autre, une société en mutation. Casablanca: Centre Jacques-Berque.
- Chaline, C. (2014). New urban policies: a geography of cities. Paris: Armand Colin.
- Chouiki, M. (2017). The Moroccan city: Between

traditions and mutations. Rabat: Publications de la Faculté des Lettres et des Sciences Humaines.
- Dekker, K. and Barthel, P. A. (2019). Urban innovations and sustainability challenges in the Maghreb. Paris : L'Harmattan.
- El Adnani, J. (2017). Local development and territorial governance in Morocco. Marrakech: Éditions Universitaires Marrakech.
- El Faiz, M. (2016). Marrakech, heritage in peril. Arles : Actes Sud.
- El Gharras, A. and Zerouali, M. (eds.) (2015). Smart cities in Morocco: Challenges and opportunities. Rabat: Royal Institute for Strategic Studies.
- Elloumi, M. and Jouve, A. M. (eds.) (2013). Bouleversements fonciers en Méditerranée : Des agricultures sous le choc de l'urbanisation et des privatisations. Paris: Karthala Editions.
- Goeury, D. and Leray, L. (2017). Urban resilience and risk management in Morocco. Rabat: Institut National d'Aménagement et d'Urbanisme.
- Harroud, T. (2018). Major urban projects in Morocco: Between global ambitions and local realities. Rabat: Éditions REMALD.
- Idrissi Janati, M. (2019). Metropolisation and territorial recompositions in Morocco. Casablanca: Éditions Afrique Orient.
- Iraki, A. and Tamim, M. (2019). Metropolitan governance in Morocco: Issues and perspectives. Casablanca: Éditions La Croisée des Chemins.
- Jaïdi, L. and Msadfa, Y. (2017). The complexity of implementing sustainable development strategies in

Morocco. Rabat: OCP Policy Center.

- Kadiri, Z. and Errahj, M. (2015). Rural leadership in Morocco: Between affirmation "from below" and recognition "from above". Alternatives Rurales, 3, 57-68.
- Khrouz, D. and Hajji, A. (eds.) (2010). North Africa in globalization: What prospects for the Maghreb? Casablanca: Konrad Adenauer Foundation.
- Lehzam, A. (2016). Urban risk governance and management in Morocco. Rabat: Éditions de l'INAU.
- Lekehal, A. (2016). Inequalities and spatial justice in Moroccan cities. Paris : Karthala.
- Mechkouri, A. and Mouloudi, H. (2019). The resilience of Moroccan coastal cities to climate change. Tangier: Publications de l'Université Abdelmalek Essaâdi.
- Naciri, M. (2017). Deserts: From yesterday to tomorrow, permanences and mutations. Rabat: Academy of the Kingdom of Morocco.
- Naciri, R. (2017). Gender and urban policies in Morocco. Rabat: Association Marocaine d'Études et de Recherches sur les Migrations.
- Navez-Bouchanine, F. (2012). Social effects of urban policies: The in-between of institutional policies and social dynamics. Paris: Karthala Editions.
- Navez-Bouchanine, F. and Berry-Chikhaoui, I. (2017). L'espace public dans les villes marocaines : Entre pratiques du quotidien et enjeux d'aménagement. Casablanca: Éditions Le Fennec.

- Niang, A. (2018). Social and solidarity economy and local development in Morocco. Rabat: Éditions REMALD.
- Oudada, M. (2018). Sustainable tourism and local development in Morocco. Agadir: Publications de l'Université Ibn Zohr.
- Peyroux, E. and Sanjuan, T. (eds.) (2016). Metropolitan strategies and city-countryside relations in Africa and Asia. Paris: Éditions Petra.
- Pinson, G. (2009). Gouverner la ville par projet: Urbanisme et gouvernance des villes européennes. Paris: Presses de Sciences Po.
- Rachik, A. (2016). Society against the state: Social movements and street strategy in Morocco. Casablanca: La Croisée des Chemins.
- Rachik, H. (2016). Moroccan society in flux: Social and cultural changes. Casablanca: La Croisée des Chemins.
- Rousset, M. and Gariel, G. (2017). Morocco in transition: From the Cherifian kingdom to the rule of law. Paris : L'Harmattan.
- Kingdom of Morocco (2018). Stratégie Nationale de Développement Urbain 2020. Rabat: Ministère de l'Aménagement du Territoire National, de l'Urbanisme, de l'Habitat et de la Politique de la Ville.
- Sedjari, A. (ed.) (2017). Governance, risks and crises. Paris: L'Harmattan.
- Semmoud, B. and Cattedra, R. (eds.). 2015. Métropolisations en Méditerranée: Cas du Maroc et de la Tunisie. Paris: Éditions Karthala.

- Taleb, A. (2017). Smart cities in Morocco: Opportunities and challenges. Casablanca: Éditions Maghrébines.
- Toulali, M. (2018). Innovation and competitiveness of territories in Morocco. Fez: Publications de l'Université Sidi Mohamed Ben Abdellah.
- Toutain, O. and Rachmuhl, V. (2014). Evaluation and impact of the Programme d'appui à la résorption de l'habitat insalubre et des bidonvilles in Morocco. Nogent-sur-Marne: GRET.
- Troin, J. F. (2015). Le Grand Maghreb (Algérie, Libye, Maroc, Mauritanie, Tunisie) : Mondialisation et construction des territoires. Paris: Armand Colin.
- Vermeren, P. (2016). Morocco in 100 questions: Un royaume de paradoxes. Paris: Tallandier.
- Zaki, L. (ed.). 2019. L'action publique au Maghreb: Enjeux professionnels et politiques. Paris : Karthala / IRMC.
- Zeino-Mahmalat, E. and Bennis, A. (eds.) (2018). Environment and climate change in the Maghreb: Challenges and prospects. Rabat: Konrad-Adenauer-Stiftung.

# *Appendices*

## Table (01): Proposed strategic plan for urban resilience in Morocco

| Strategic focus | Objectives | Concrete actions | Performance indicators |
|---|---|---|---|
| 1. Governance and resilient urban planning | 1.1 Strengthening multi-level coordination | - Create an interministerial committee for urban resilience <br> - Developing local resilience plans | - Number of committee meetings <br> - of towns with a local resilience plan |
| | 1.2 Integrating resilience into urban planning documents | - Revise SDAUs and PAUs to include resilience criteria <br> - Training urban planners in resilience issues | - % of urban planning documents integrating resilience <br> - Number of planners trained |
| 2. Inclusive economic development | 2.1 Diversifying the local economy | - Creating sector-based competitiveness clusters <br> - Supporting entrepreneurship and innovation | - Number of new economic sectors developed <br> - Business start-up rate |
| | 2.2 Promoting the social economy | - Setting up SSE incubators <br> - Facilitating access to financing for cooperatives | - Number of SSE structures created <br> - Amount of funding allocated to the SSE |
| 3. Social cohesion and reducing inequalities | 3.1 Improving access to decent housing | - Developing social housing programs <br> - Rehabilitating informal settlements | - Number of social housing units built <br> - % of population living in rehabilitated neighborhoods |

| | | | |
|---|---|---|---|
| - | 3.2 Strengthening citizen participation | - Setting up participatory budgets<br>- Creating neighborhood councils | - % of municipal budget allocated to participatory projects<br>- Neighborhood council participation rates |
| 4. Adapting to climate change | 4.1 Reducing vulnerability to natural hazards | - Mapping risk areas<br>- Set up early warning systems | - % of mapped territory<br>- Alert reaction time |
| | 4.2 Promoting energy efficiency | - Energy-efficient renovation of public buildings -Developing urban renewable energies | - % reduction in energy consumption<br>- Share of renewable energies in the urban energy mix |
| 5. Sustainable resource management | 5.1 Optimizing water management | - Upgrading drinking water networks<br>- Promoting the reuse of treated wastewater | - Water loss reduction rate<br>- Volume of reused water |
| - | 5.2 Improving waste management | - Develop selective sorting and recycling<br>- Creating waste recovery channels | - Recycling rate<br>- Quantity of waste recycled |
| 6. Sustainable urban mobility | 6.1 Developing public transport | - Create BRT or tramway lines<br>- Improving intermodality | - Km of public transport lines created Modal share of public transport |

| | 6.2 Promoting soft mobility | - Developing bicycle paths<br>- Pedestrianize city centers | - Km of cycle paths<br>- Pedestrianized area |
|---|---|---|---|
| 7. Capacity building and innovation | 7.1 Training local players | - Creating an academy of urban resilience<br>- Organize exchanges of experience between cities | - Number of actors trained<br>- Number of exchanges organized |
| | 7.2 Encouraging urban innovation | - Call for innovative projects<br>- Creating urban living labs | - Number of innovative projects financed<br>- Number of innovations tested |

# Table (02): Operational strategy for adapting Moroccan cities to the challenges of globalization

| Strategic focus | Objectives | Operational actions | Players involved | Follow-up indicators |
|---|---|---|---|---|
| 1. Economic competitiveness | 1.1 Attracting foreign direct investment | - Creating urban tax-free zones<br>- Simplifying administrative procedures for investors<br>- Developing a territorial marketing strategy | - Ministry of Industry<br>-CRI<br>- Local authorities | - Number of FDI attracted<br>- Amount of investments made |
| | 1.2 Developing high value-added sectors | - Set up sector-based clusters (IT, aeronautics, automotive)<br>- Create urban technology parks<br>- Supporting R&D and innovation | - Ministry of Higher Education<br>- CNRST<br>- Private companies | - Number of jobs created in targeted sectors<br>- Number of patents filed |
| 2. Connectivity and infrastructure | 2.1 Improving digital connectivity | - Deploying 5G in major cities<br>- Developing fiber optic networks | -ANRT<br>- Telecom operators<br>- Local authorities | - Rate of 5G coverage<br>- % of population with access to very high-speed broadband |

| | | - Creating connected public spaces | | |
|---|---|---|---|---|
| 3. Human capital and training | 2.2 Modernizing transport infrastructures | - Developing multimodal logistics platforms<br>- Improve connections between ports, airports and urban areas | - Ministry of Public Works - ONCF<br>- ADM | - Lower logistics costs<br>- Transit times between economic hubs |
| | | - Setting up intelligent transport systems | | |
| | 3.1 Matching skills to business needs global market | - Creating university-business partnerships<br>- Developing foreign language training courses<br>- Set up programs continuing education | - Ministry of Education<br>- OFPPT<br>- Chambers of commerce | - Graduates' professional integration rate<br>- Number of training courses created in partnership with companies |

| | 3.2 Attracting and retaining talent | - Creating incubators and nurseries companies<br>- Set up programs mentoring | - ANAPEC<br>- Start-up Morocco<br>- Diaspora Moroccan | - Number of startups created<br>- Return rate talents Moroccans from Abroad |
| --- | --- | --- | --- | --- |

# I want morebooks!

Buy your books fast and straightforward online - at one of world's fastest growing online book stores! Environmentally sound due to Print-on-Demand technologies.

Buy your books online at
## www.morebooks.shop

Kaufen Sie Ihre Bücher schnell und unkompliziert online – auf einer der am schnellsten wachsenden Buchhandelsplattformen weltweit! Dank Print-On-Demand umwelt- und ressourcenschonend produziert.

Bücher schneller online kaufen
## www.morebooks.shop

info@omniscriptum.com
www.omniscriptum.com

Printed by Books on Demand GmbH, Norderstedt / Germany